# Lecture Notes in Economics and Mathematical Systems

Managing Editors: M. Beckmann and H. P. Künzi

Systems Theory



115

B. D. O. Anderson · M. A. Arbib
E. G. Manes



# Foundations of System Theory: Finitary and Infinitary Conditions



Springer-Verlag
Berlin · Heidelberg · New York 1976

**Authors**
Brian D. O. Anderson
Department of Electrical Engineering
University of Newcastle
New South Wales 2308/Australia

Michael A. Arbib
Department of Computer and Information Science
University of Massachusetts
Amherst, Massachusetts, 01002/USA

Ernest G. Manes
Department of Mathematics
University of Massachusetts
Amherst, Massachusetts, 01002/USA

**Library of Congress Cataloging in Publication Data**

Arbib, Michael A
Foundations of system theory.

(Lecture notes in economics and mathematical systems ; 115)
Bibliography: p.
Includes index.
1. System theory. I. Manes, Ernest G., 1943- joint author. II. Anderson, Brian D. O., joint author. III. Title. IV. Series.
Q295.A7 003 76-1967

AMS Subject Classifications (1970): 18B20, 18C10, 93A99, 93B15

ISBN 3-540-07611-5 Springer-Verlag Berlin · Heidelberg · New York
ISBN 0-387-07611-5 Springer-Verlag New York · Heidelberg · Berlin


Printed in Germany
Printing and binding: Beltz Offsetdruck, Hemsbach/Bergstr.

## INTRODUCTION

This paper is one of a series in which the ideas of category theory are applied to problems of system theory. As with the three principal earlier papers, [1-3], the emphasis is on study of the realization problem, or the problem of associating with an input-output description of a system an internal description with something analogous to a state-space.

In this paper, several sorts of machines will be discussed, which arrange themselves in the following hierarchy:

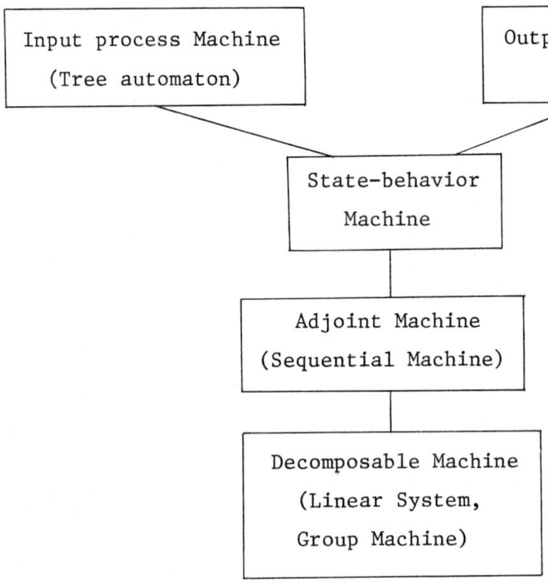


Each member of the hierarchy includes members below it; examples are included in parentheses, and each example is at its lowest possible point in the hierarchy. There are contrived examples of output process machines and state-behavior machines which are not adjoint machines [3], but as yet, no examples with the accepted stature of linear systems [4], group machines [5, 6], sequential machines [7, Ch. 2], and tree automata [7, Ch. 4].

To grasp in very great generality what we attempt to do in this paper, we recall several facts concerning discrete-time linear systems:

(1) One can take an external description of a linear system, in the form of a map $f^{\blacktriangle}$ taking past input sequences into future output sequences and, using results of module theory, construct an internal realization essentially by factoring $f^{\blacktriangle}$ into an onto linear map followed by a one-to-one linear map [4]. The codomain of the onto linear map is the state-space, and is (Space of Input Sequences)/Ker $f^{\blacktriangle}$. (It is actually possible to do all this in matrix terms, working with Hankel matrices of Markov parameters, and this viewpoint may be more familiar to some.)

(2) One can use a Nerode equivalence class theory [8], originally developed for sequential machines, to obtain the set of reachable and observable states associated with a map $f^{\blacktriangle}$. Then one can observe that the set is really a linear space, and the procedure is equivalent to (1).

(3) In a finite-dimensional reachable and observable linear system, all states can be reached in a bounded time, and all can be observed in a bounded time [9].

In [2], we extended (1) by recalling from category theory the concept of $\mathcal{E}$-$\mathcal{M}$ factorization which generalizes the idea of factoring into an onto or epi map followed by a one-to-one or mono map.

In this paper, we attempt to expand the application of the other ideas to various classes of machines. To extend (2), we attempt to force a Nerode equivalence structure into the more general situations. To extend (3) and obtain a notion of finiteness in a more general context, we generalize the idea of reachable (and observable) in a finite time.

As background, we require familiarity with [2], but not necessarily [1] or [3], though we should be dishonest were we to claim that knowledge of [1] and [3] would be of no extra help. Likewise, though knowledge of some category theory, as per say [10, 11, or 12], would be helpful, we only strictly require knowledge of those category theory ideas used in [2], introducing other category theory concepts as required.

The body of the paper is organized into a structure best depicted as follows:

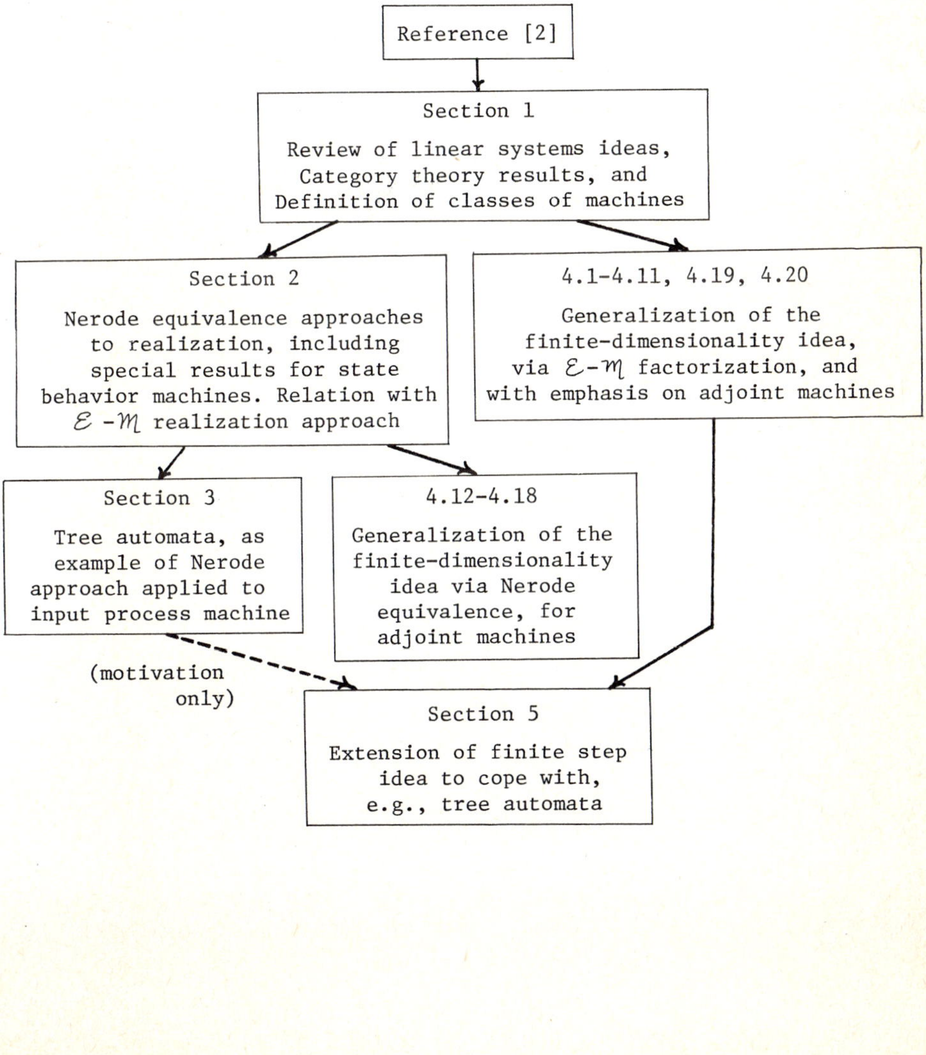

As the diagram shows, the paper in its totality is very much a sequel to [2].

Certainly, it is clear that the ideas of this paper unify a number of apparently distinct ideas scattered in the literature. Unification is not all, however: by proving results about linear systems in a category theory rather than vector space setting, one strips away features of the vector space setting which may obscure possible applications or extensions of the result in other settings.

Acknowledgement:

The research reported in this volume was supported in part by National Science Foundation grant number GJ 35759, which also supported Dr. Anderson's visit to the University of Massachusetts at Amherst for the period September 1973 through February 1974. This work was completed in February 1974.

TABLE OF CONTENTS

## 1. A General Setting for Discrete Action Nonlinear Systems.

In the paper "Foundations of System Theory: Decomposable Systems", [2], to which this is a sequel, we showed how the techniques of category theory allowed us to give a general theory of reachability, observability, realization and duality for decomposable systems: a class embracing both linear (discrete-time) systems and group machines. In "Machines in a Category: An Expository Introduction", [1], we gave a more general categorical system theory which handles nonlinear systems as well as decomposable systems. In fact, as we shall see in Section 3, it also handles the tree automata of formal language theory and computer semantics, so long as certain finiteness conditions are met.

In this section, we assume the reader familiar with the notions of [2] --such as category, power and copower, dynamorphism of decomposable dynamics, etc.--and provide the extra background from category theory required to set forth the general framework of [1].

The time-invariant system

$$\dot{q} = f(q,u) \qquad y = \beta(q) \tag{1}$$

when discretized in time (into a difference equation), initialized (coordinates are chosen such that $f(0,0) = 0$, $\beta(0) = 0$), and linearized, gives rise to the linear system

$$q(t + 1) = Fq(t) + Gu(t) \qquad y(t) = Hq(t) \tag{2}$$

described by the linear mappings $F: Q \rightarrow Q$, $G: U \rightarrow Q$, $H: Q \rightarrow Y$ of finite-dimensional real vector spaces. Similarly, the time-varying system

$$\dot{q} = f(q,u,t) \qquad y = \beta(q,t) \tag{3}$$

when discretized, initialized and linearized is described by the sequences

$$F_t: Q_{t-1} \longrightarrow Q_t, \quad G_t: U \longrightarrow Q_t, \quad H_t: Q_t \longrightarrow Y \qquad (t \in \underline{\underline{R}}) \qquad (4)$$

of linear mappings of finite-dimensional vector spaces. Discretizing (1) without initialization or linearization produces the nonlinear system

$$q(t+1) = \delta(q(t), u(t)) \qquad y(t) = \beta(q(t)) \qquad (5)$$
$$q(0) = q_0$$

where the functions $\delta: Q \times U \longrightarrow Q$ and $\beta: Q \longrightarrow Y$ are not required to satisfy linearity conditions.

The theory of [2] is designed to study a class of machines which contains, but also extends beyond, the linear systems of (2). That of [1] on the other hand is applicable in a wider variety of situations, including the nonlinear systems of (5).

## DECOMPOSABLE MACHINES

We first recall the key definitions and results from [2] (where they were numbered 2.9, 2.13, 2.16, 2.20).

**1.1 DEFINITION**: Let $\mathcal{K}$ be any category. Then, a system dynamics in $\mathcal{K}$ is a pair (Q,F) where Q is an object of $\mathcal{K}$ and $F: Q \longrightarrow Q$ is a $\mathcal{K}$-morphism. A morphism of dynamics $g: (Q,F) \longrightarrow (Q',F')$, called a dynamorphism, is a $\mathcal{K}$-morphism $g: Q \longrightarrow Q'$ such that $F' \cdot g = g \cdot F$ as shown below:

$$\begin{array}{ccc} Q & \xrightarrow{\ g\ } & Q' \\ {\scriptstyle F}\downarrow & & \downarrow{\scriptstyle F'} \\ Q & \xrightarrow[\ g\ ]{} & Q' \end{array}$$

It is obvious that the identity map $id_Q: (Q,F) \longrightarrow (Q,F)$ is a dynamorphism and that the composition $gf: (Q_1,F_1) \longrightarrow (Q_3,F_3)$ of the dynamorphism

$f: (Q_1, F_1) \to (Q_2, F_2)$ and the dynamorphism $g: (Q_2, F_2) \to (Q_3, F_3)$ is again a dynamorphism $[(gf)F_1 = g(fF_1) = g(F_2 f) = (gF_2)f = (F_3 g)f = F_3(gf)]$. This gives rise to the <u>category of system dynamics</u> in $\mathcal{K}$ which we denote $\text{Dyn}(\mathcal{K})$, whose objects are system dynamics, and whose morphisms are dynamorphisms.

A <u>decomposable system in</u> $\mathcal{K}$ is a 6-tuple $M = (Q, F, I, Y, G, H)$ such that $(Q, F)$ is a system dynamics in $\mathcal{K}$, $G$ is a $\mathcal{K}$-morphism of the form $G: I \to Q$ (the <u>input map</u>) and $H$ is a $\mathcal{K}$-morphism of the form $H: Q \to Y$ (the <u>output map</u>).

**1.2 LEMMA:** Let the $\mathcal{K}$-object $Y$ have countable power $(\pi_k: Y_\S \to Y \mid k \geq 0)$ and define the 'left shift' $z_Y: Y_\S \to Y_\S$ by $\pi_k z = \pi_{k+1}$, $k \geq 0$. Given any pair of $\mathcal{K}$-morphisms $Q \xrightarrow{H} Y$ and $Q \xrightarrow{F} Q$ there exists a unique $\mathcal{K}$-morphism $\sigma: Q \to Y_\S$ such that the diagram

$$Y \xleftarrow{\pi_0} Y_\S \xleftarrow{z_Y} Y_\S$$
$$\sigma \uparrow \qquad \sigma \uparrow \qquad H \nearrow$$
$$Q \xleftarrow{F} Q \qquad\qquad\qquad (6)$$

commutes, namely that satisfying the equation $\pi_k \sigma = HF^k$. □

**1.3 LEMMA:** Let the $\mathcal{K}$-object $I$ have countable copower $(\text{in}_j: I \to I^\S \mid j \geq 0)$ and define the 'left shift' $z_I$ by $z \cdot \text{in}_j = \text{in}_{j+1}$, $j \geq 0$. Given any pair of $\mathcal{K}$-morphisms $I \xrightarrow{G} Q$ and $Q \xrightarrow{F} Q$ there exists a unique $\mathcal{K}$-morphism $r: I^\S \to Q$ such that the diagram

$$I \xrightarrow{\text{in}_0} I^\S \xrightarrow{z_I} I^\S$$
$$G \searrow \quad \downarrow r \qquad \downarrow r \qquad\qquad (7)$$
$$Q \xrightarrow{F} Q$$

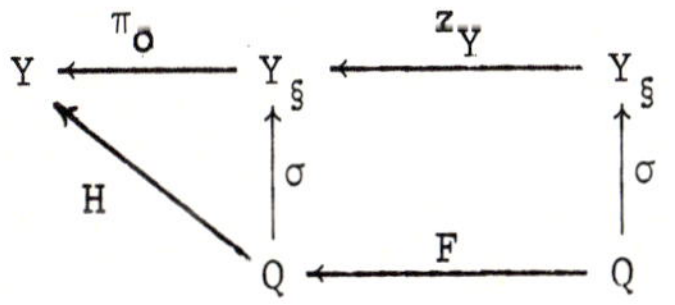

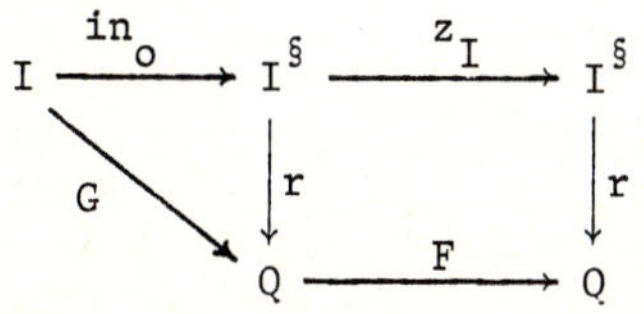

commutes, namely that satisfying the equation $r \cdot in_j = F^j G$. □

1.4 DEFINITION: Let $M = (Q,F,I,G,Y,H)$ be a system in the category $\mathcal{K}$ with countable powers and copowers. Then M has reachability map

$$r: I^{\S} \rightarrow Q \qquad \text{defined by } r \cdot in_j = F^j G$$

and observability map

$$\sigma: Q \rightarrow Y_{\S} \qquad \text{defined by } \pi_k \cdot \sigma = HF^k.$$

The total response of M is then the composition

$$f^{\blacktriangle} = \sigma \cdot r: I^{\S} \rightarrow Y_{\S}.$$

The response map of M is simply

$$f = \pi_0 \cdot f^{\blacktriangle}: I^{\S} \rightarrow Y.$$

As in the case of linear systems, r, $\sigma$ and $f^{\blacktriangle}$ (but not f) are all dynamorphisms, as we see by reading off the squares of diagrams (6) and (7), and noting that the composite of dynamorphisms is again a dynamorphism.

## MACHINES IN A CATEGORY

To relate these concepts to the general framework of [1], we need the concepts of functor and of adjoint from category theory. (For further background on standard category theory see the textbooks of Arbib and Manes [12], Herrlich and Strecker [11] and Mac Lane [10].) The definitions of left and right adjoint will seem rather indigestible at first, but should become clearer when we rework Lemmas 1.2 and 1.3 into the forms 1.12 and 1.11 below:

1.5 DEFINITION: Let $\mathcal{K}$ and $\mathcal{L}$ be categories. Then a functor $H: \mathcal{K} \rightarrow \mathcal{L}$ comprises a map $A \mapsto AH$ which sends objects A of $\mathcal{K}$ to objects AH of $\mathcal{L}$, together with, for each pair A,B of objects of $\mathcal{K}$, a map

$$\mathcal{K}(A,B) \to \mathcal{L}(AH,BH): f \mapsto fH$$

which sends $\mathcal{K}$-morphisms $f: A \to B$ to $\mathcal{L}$-morphisms $fH: AH \to BH$ in such a way that

(i) $id_A H = id_{AH}$ for every object A of $\mathcal{K}$

(ii) $(f\cdot g)H = fH\cdot gH$ for every pair $f,g$ of composable $\mathcal{K}$-morphisms.

1.6 EXAMPLES:

(i) For any category $\mathcal{K}$, the assignment $A \mapsto A$, $f \mapsto f$ clearly defines a functor--1.5(i) and (ii) are trivally satisfied--called the identity functor $id_{\mathcal{K}}$ of $\mathcal{K}$.

(ii) In the category Set we may fix on any set $X_0$ to form the assignment $A \mapsto A\times X_0$; $(f: A \to B) \mapsto (f\times X_0: A\times X_0 \to B\times X_0: (a,x) \mapsto (f(a),x))$.

This defines a functor since

$$id_A\times X_0 = id_{A\times X_0}: (a,x) \mapsto (a,x)$$

and $(f\cdot g)\times X_0 = (f\times X_0)\cdot(g\times X_0): (a,x) \mapsto (f(g(a)),x)$.

We call this functor $-\times X_0$: Set $\to$ Set.

This notion of functor allows us to give one of the key definitions of [1]. Below, we shall relate it to the just reviewed ideas of [2].

1.7 DEFINITIONS:

(i) A process X in a category $\mathcal{K}$ is a functor $X: \mathcal{K} \to \mathcal{K}$.

(ii) An X-dynamics is a pair $(Q,\delta)$ with Q an object (the state object) of $\mathcal{K}$ and $\delta: QX \to Q$ a morphism of $\mathcal{K}$.

(iii) An I-frame† is a pair $(I,\tau)$ with I an object (the input object) and $\tau: I \to Q$ a morphism

†This rather neutral terminology is due to the fact that for decomposable machines we interpret $(I,\tau)$ as specifying the Input map, while for non-linear systems, we think of it as giving the Initial state.

(iv) An output map is a pair $(Y,\beta)$ with Y an object (the output object) and $\beta: Q \to Y$ a morphism

(v) An X-system is a 7-tuple $(X,Q,\delta,I,\tau,Y,\beta)$ such that X is a process, $(Q,\delta)$ an X-dynamics, $(I,\tau)$ an I-frame, and $(Y,\beta)$ an output map.

1.8 EXAMPLES:

(i) Let $\mathcal{K}$ = Vect, the category of real vector spaces and linear maps. Let the functor $X: \mathrm{Vect} \to \mathrm{Vect}$ be the identity functor, $QX = Q$, $fX = f$. Then (2) is captured as the X-dynamics $(Q,F)$ with I-frame the input map $(U,G)$ and output map $(Y,H)$. These are the decomposable systems of 1.1 above.

(ii) Let $\mathcal{K}$ be the category whose objects are sequences $(Q_t \mid t \in \mathbb{Z})$ of real vector spaces and let the $\mathcal{K}$-morphisms $(Q_t) \to (R_t)$ be sequences $(f_t \mid t \in \mathbb{Z})$ where each $f_t: Q_t \to R_t$ is linear. Identities and composition are the ordinary ones "t-wise". Let the functor $X: \mathcal{K} \to \mathcal{K}$ be defined by $(Q_t)X = (\bar{Q}_t)$ with $\bar{Q}_t = Q_{t-1}$ and, for $(f_t): (Q_t) \to (R_t)$, $(f_t)X = (\bar{f}_t)$ with $\bar{f}_t: \bar{Q}_t \to \bar{R}_t = f_{t-1}$. Then (4) is captured by the X-dynamics $((Q_t),(F_t))$, with I-frame $((I),(G_t))$ and output map $((Y),(H_t))$. Here, $(I)$ is the constant sequence $(I_t)$ with $I_t = I$, and $(Y)$ similarly.

(iii) Let $\mathcal{K}$ be the category Set of sets and functions. If $X_o$ is a fixed set define $X: \mathrm{Set} \to \mathrm{Set}$ by $QX = Q \times X_o$ and, for $f: Q \to R$,

$$fX = f \times X_o : Q \times X_o \to R \times X_o, \qquad (q,x) \to (f(q),x).$$

Then (5) is described by the X-dynamics $(Q,\delta)$, with I-frame the initial state $(I,q_o)$ (where I is a one-element set), and output map $(Y,\beta)$. [Here, the following trick is being used: one can identify one particular element c of an arbitrary set C with one particular mapping $\gamma\colon I \to C$ where I is a one-element set, since $\gamma$ is uniquely specified by its image.] The sextuple $(X_o, Q, \delta, q_o, Y, \beta)$ is the well-known sequential machine [ 7].

Other examples will be given in the course of the paper.

Next, following [1], we introduce the concepts that allow X-systems to "run", i.e. which allow us to set up morphisms that in some way parallel the reachability morphism of decomposable machines, which, for linear systems, relates past input sequences (as opposed to single inputs) to the present state.

<u>1.9</u> <u>DEFINITION</u>: Let $X\colon \mathcal{K} \to \mathcal{K}$ be a process in $\mathcal{K}$. The <u>category of X-dynamics</u>, written Dyn(X) is defined to be the category whose objects are X-dynamics $(Q,\delta)$ and whose morphisms are X-<u>dynamorphisms</u> $\psi\colon (Q,\delta) \to (Q',\delta')$, being $\mathcal{K}$-morphisms $\psi\colon Q \to Q'$ satisfying

$$\begin{array}{ccc} QX & \xrightarrow{\ \delta\ } & Q \\ {\scriptstyle \psi X}\downarrow & & \downarrow{\scriptstyle \psi} \\ Q'X & \xrightarrow{\ \delta'\ } & Q' \end{array} \qquad (8)$$

Notice that $id_Q\colon (Q,\delta) \to (Q,\delta)$ is a dynamorphism and $\psi_2\psi_1\colon (Q,\delta) \to (Q'',\delta'')$ is a dynamorphism if $\psi_1\colon (Q,\delta) \to (Q',\delta')$ and $\psi_2\colon (Q',\delta') \to (Q'',\delta'')$ are; this ensures that Dyn(X) is indeed a category.

<u>1.10</u> <u>OBSERVATION</u>: "Forgetting the dynamics" which sends an X-dynamics $(Q,\delta)$ to the state object Q, and which views a dynamorphism $\psi\colon (Q,\delta) \to (Q',\delta')$ simply as a $\mathcal{K}$-morphism $\psi\colon Q \to Q'$ defines a functor $U\colon \mathrm{Dyn}(X) \to \mathcal{K}$ --we call it the <u>forgetful functor</u>.

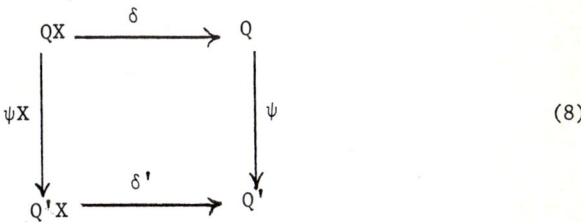

With this, we may recall from [1]:

**1.11 DEFINITION:** X is an *input process* if the forgetful functor $\mathrm{Dyn}(X) \to \mathcal{K}: (Q,\delta) \mapsto Q$ has a *left adjoint* (left adjoint is a standard category theory notion) i.e., if for each object I of $\mathcal{K}$ there exists an object $IX^{@}$ with two associated morphisms, $\eta: I \to IX^{@}$ and a *free dynamics* $\mu_o: (IX^{@})X \to IX^{@}$; which are such that, for any X-dynamics $(Q,\delta)$ and any $\mathcal{K}$-morphism $h: I \to Q$, there exists a unique dynamorphism $h^{\#}: (IX^{@},\mu_o) \to (Q,\delta)$ such that $h^{\#}\cdot\eta = h$ :

$$\begin{array}{ccc} I & \xrightarrow{\eta} & IX^{@} \\ & {\scriptstyle h}\searrow & \downarrow{\scriptstyle h^{\#}} \\ & & Q \end{array} \qquad \begin{array}{ccc} (IX^{@})X & \xrightarrow{\mu_o} & IX^{@} \\ \downarrow{\scriptstyle h^{\#}X} & & \downarrow{\scriptstyle h^{\#}} \\ QX & \xrightarrow{\delta} & Q \end{array} \tag{9}$$

We call $h^{\#}$ the *dynamorphic extension* of h.

Several remarks should immediately be made.

(i) The above definition may be employed in a situation where (I,h) is thought of as an I-frame g but this is not always so.

(ii) To make contact with other viewpoints of left adjointness which the reader may have, observe that the above definition associates with any $h \in \mathcal{K}[I,\mathcal{K}[I,Q]]$ (for any I and $(Q,\delta)$), a unique $h^{\#}$ in $[(IX^{@},\mu_o),(Q,\delta)]$. It is also immediate from the diagram that any $h^{\#} \in \mathrm{Dyn}(X)[(IX^{@},\mu_o),(Q,\delta)]$ uniquely determines an h by $h^{\#}\eta = h$.

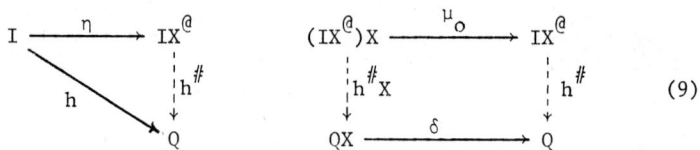

(iii) Observe that the free dynamics is uniquely determined to within isomorphism by I. For if $\tilde{\mu}_o: RX \to R$, $\tilde{\eta}: I \to R$ were another model of a free dynamics then one could form $\tilde{\eta}^{\#}$ by the freeness of $IX^{@}$ and $\eta^{\#}$ by the freeness of R such that

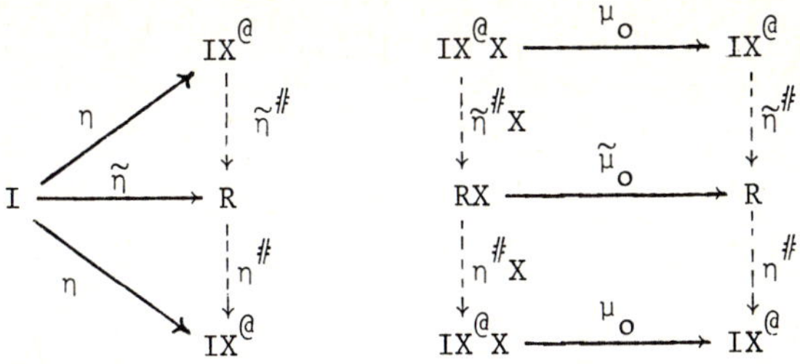


Now observe that $h^{\#} = \eta^{\#}\tilde{\eta}^{\#}$ satisfies $h^{\#}\eta = \eta$ and is a dynamorphism. The same is true of $id_{IX^{@}}$. Noting the uniqueness of $h^{\#}$ in the definition above, it follows that $\eta^{\#}\tilde{\eta}^{\#} = id$. Similarly, one can show $\tilde{\eta}^{\#}\eta^{\#} = id$. Hence $(R,\tilde{\mu}_o)$ and $(IX^{@},\mu_o)$ are isomorphic. (This result, when viewed as a theorem about adjoints, is standard.)

To see that this generalizes 1.3, we may immediately note:

**1.12 FACT**: If each object of $\mathcal{K}$ has countable copowers, then $id_{\mathcal{K}}$ is an input process.

Proof: If one inserts in (9)

$I^{\S}$ for $IX^{@}$

$in_o: I \longrightarrow I^{\S}$ for $\eta: I \longrightarrow IX^{@}$

$z_I: I^{\S} \longrightarrow I^{\S}$ for $\mu_o: (IX^{@})X \longrightarrow IX^{@}$

the fact that F and G determine a unique r in (7) becomes equivalent to the fact that h (= G) and $\delta$ (= F) determine a unique $h^{\#}$ (= r) in (9). □

"Reversing the arrows" in 1.11, we may recall the following definition from [3]. (The notion of right adjoint is standard category theory, and will be taken up again at 1.23 below.)

1.13 DEFINITION: X is an output process if the forgetful functor $\mathrm{Dyn}(X) \to \mathcal{K}\colon (Q,\delta) \to Q$ has a right adjoint; i.e. if for each $Y \in \mathcal{K}$ there exists an object $YX_@$ with two associated morphisms, $\Lambda\colon YX_@ \to Y$ and a cofree dynamics $L\colon (YX_@)X \to YX_@$; with the property that, given any X-dynamics $(Q,\delta)$ and any $\mathcal{K}$-morphism $h\colon Q \to Y$, there exists a unique dynamorphism $h_{\#}\colon (Q,\delta) \to (YX_@,L)$ such that $\Lambda \cdot h_{\#} = h$ :

$$\begin{array}{ccc} Y & \xleftarrow{\ \Lambda\ } & YX_@ \\ & \nwarrow^{h} & \uparrow h_{\#} \\ & & Q \end{array} \qquad\qquad \begin{array}{ccc} (YX_@)X & \xrightarrow{\ L\ } & YX_@ \\ \uparrow h_{\#}X & & \uparrow h_{\#} \\ QX & \xrightarrow{\ \delta\ } & Q \end{array} \tag{10}$$

For the moment, the choice of object names as Q and Y has no special significance.

Just as for free dynamics, so can the cofree dynamics be proved unique up to isomorphism.

We see now that 1.2 yields

1.14 FACT: If each object of $\mathcal{K}$ has countable powers, then $id_{\mathcal{K}}$ is an output process, with

$$YX_@ = Y_§ \quad \text{for each } \mathcal{K}\text{-object } Y$$

while $$YX_@ \xrightarrow{\ \Lambda\ } Q = Y_§ \xrightarrow{\ \pi_0\ } Y$$

and $$(YX_@)X \xrightarrow{\ L\ } YX_@ = Y_§ \xrightarrow{\ z_Y\ } Y_§ \ .$$ □

1.15 DEFINITION: $X\colon \mathcal{K} \to \mathcal{K}$ is a state-behavior process if it is both an input process and an output process.

The decomposable machines of [2], including of course linear systems, have, by 1.12 and 1.14 processes which are state-behavior. The theory also embraces that of sequential machines:

<u>1.16</u> <u>FACT</u>: The process $-\times X_o$: <u>Set</u> $\longrightarrow$ <u>Set</u> is state-behavior.

Proof: Given a set $X_o$, let $X_o^*$ be the set of all finite strings $(x_1,\ldots,x_n)$ of elements of $X_o$ --we write $\Lambda$ for the empty string ( ), and distinguish an element x of $X_o$ from the corresponding string (x) of $X_o^*$.

We must first check that $X = -\times X_o$ is an input process. We claim that the following definitions do the trick:

$$IX^@ = I\times X_o^* \quad \text{for each set } I$$

$$\eta: I \longrightarrow IX^@ = I \longrightarrow I\times X_o^*: i \longmapsto (i,\Lambda)$$

$$\mu_o: (IX^@)X \longrightarrow IX^@ = (I\times X_o^*)\times X_o \longrightarrow I\times X_o^*: ((i,w),x) \longmapsto (i,wx)$$

Here, $wx = (x_1,\ldots,x_n,x)$ for $w = (x_1,\ldots,x_n) \in X_o^*$ and $x \in X_o$.

What we must check is that, in terms of the notation of <u>1.11</u>, an arbitrary $h: I \longrightarrow Q$ gives rise to a unique $h^{\#}$ with (from the left diagram)

$$h^{\#}(i,\Lambda) = h(i) \quad \text{for each } i \in I$$

and (from the right diagram)

$$h^{\#}(i,wx) = \delta(h^{\#}(i,w),x) \quad \text{for each } i \in I,\ w \in X_o^*,\ x \in X .$$

In fact, these uniquely define $h^{\#}$ by the rule

$$h^{\#}(i,w) = \delta^*(h(i),w)$$

where $\delta^*$ is the standard extension of $\delta$ from $Q\times X_o \longrightarrow Q$ to $Q\times X_o^* \longrightarrow Q$ via the induction rule $\delta^*(q,\Lambda) = q$ and $\delta^*(q,wx) = \delta(\delta^*(q,w),x)$.

We next check that $X = -\times X_o$ is an output process with the formulae

$$YX_@ = Y^{X_o^*} \quad \text{the set of all maps from } X_o^* \text{ to } Y$$

$$\Lambda: YX_@ \longrightarrow Y = Y^{X_o^*} \longrightarrow Y: f \longmapsto f(\Lambda)$$

$$L: (YX_@)X \longrightarrow YX_@ = Y^{X_o^*}\times X_o \longrightarrow Y^{X_o^*}: (f,x) \longmapsto fL_x$$

where $fL_x$ denotes composition of $f$ with "the left juxtaposition of $x$" map, $L_x: X_o^* \to X_o^*: w \mapsto xw$. Thus $fL_x(w) = f(xw)$. The diagrams of <u>1.13</u> require that, given an arbitrary $h: Q \to Y$, there be $h_\#$ with

$$[h_\#(q)](\Lambda) = h(q) \quad \text{for each } q \in Q \quad (\text{recall that } h_\#(q) \text{ is a map } X_o^* \to Y)$$

and

$$h_\#(\delta(q,x)) = h_\#(q)L_x \quad \text{for each } q \in Q, \; x \in X_o$$

i.e. $\quad [h_\#(\delta(q,x))](w) = [h_\#(q)L_x](w) = [h_\#(q)](xw)$

for each $q \in Q$, $x \in X_o$ and $w \in X_o^*$. A routine calculation then verifies that $h_\#(q)$ is uniquely determined to be

$$h_\#(q): X_o^* \to Y: w \mapsto h[\delta^*(q,w)] \;. \qquad \square$$

<u>**1.17**</u> **DEFINITION**: Let $M = (X,Q,\delta,I,\tau,Y,\beta)$ be an X-system. If $X$ is an input process, the <u>reachability map</u> $r: IX^@ \to Q$ of $M$ is the unique dynamorphism $r: (IX^@,\mu_o) \to (Q,\delta)$ which satisfies

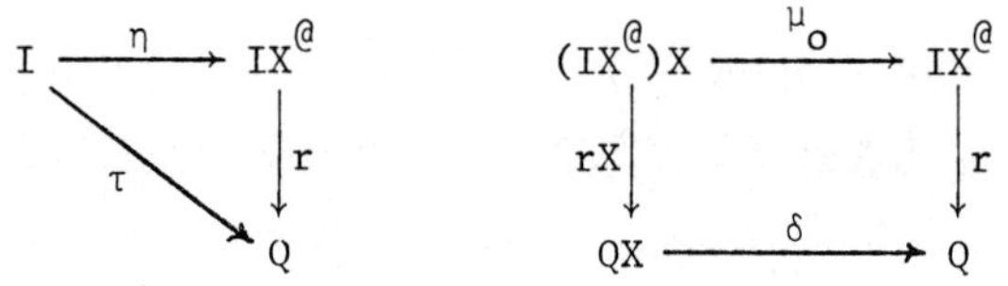

$$
\begin{array}{cc}
\begin{array}{c}
I \xrightarrow{\eta} IX^@ \\
{}_\tau \searrow \quad \downarrow r \\
\quad\quad Q
\end{array}
&
\begin{array}{c}
(IX^@)X \xrightarrow{\mu_o} IX^@ \\
rX \downarrow \quad\quad\quad \downarrow r \\
QX \xrightarrow{\delta} Q
\end{array}
\end{array} \tag{11}
$$

If $X$ is an output process, the <u>observability map</u> $\sigma: Q \to YX_@$ of $M$ is the unique dynamorphism $\sigma: (Q,\delta) \to (YX_@, L)$ which satisfies

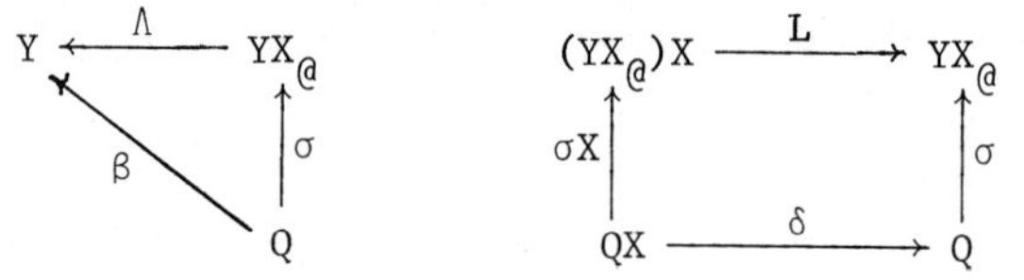

$$
\begin{array}{cc}
\begin{array}{c}
Y \xleftarrow{\Lambda} YX_@ \\
{}_\beta \nwarrow \quad \uparrow \sigma \\
\quad\quad Q
\end{array}
&
\begin{array}{c}
(YX_@)X \xrightarrow{L} YX_@ \\
\sigma X \uparrow \quad\quad\quad \uparrow \sigma \\
QX \xrightarrow{\delta} Q
\end{array}
\end{array} \tag{12}
$$

If $X$ is a state-behavior process, the total response $f^\blacktriangle: IX^@ \to YX_@$ of $M$ is then the composite dynamorphism $f^\blacktriangle = \sigma \cdot r: (IX^@,\mu_o) \to (YX_@,L)$.

The <u>response map</u> (or <u>system behavior</u>) of $M$ is simply

$$f = \Lambda \cdot f^{\blacktriangle} = \beta \cdot r \colon IX^{@} \longrightarrow Y$$

and is not a dynamorphism.

Note that, by 1.12 and 1.14, these definitions subsume those for decomposable machines in 1.4. They also include many other situations, for example, the sequential machine case.

1.18 EXAMPLE: An X-system $(X,Q,\delta,I,\tau,Y,\beta)$ becomes a sequential machine $(X_o,Q,\delta,q_o,Y,\beta)$ in the sense of 1.8 example (iii), if we take $\mathcal{K}$ to be Set, $X$ to be the functor $-\times X_o$; $Q$ to be a set; $\delta$ a map $Q\times X_o \longrightarrow Q$; $I$ to be a one element set so that $\tau\colon I \longrightarrow Q$ becomes an element $q_o \in Q$ and $I\times X_o^* \cong X_o^*$ ; $Y$ to be a set and $\beta\colon Q \longrightarrow Y$. Using the formulae obtained in 1.16, we see immediately that such a machine M has

reachability map $r\colon X_o^* \longrightarrow Q : w \mapsto \delta^*(q_o,w)$

sending w to the state M reaches after starting in its initial state $q_o$ and receiving input string w;

observability map $\sigma\colon Q \longrightarrow Y^{X_o^*} : q \mapsto M_q$

where $M_q\colon X_o^* \longrightarrow Y : w \mapsto \beta[\delta^*(q,w)]$ tells us the output of M started in state q which will be observed after feeding it any input string w;

total response $f^{\blacktriangle}\colon X_o^* \longrightarrow Y^{X_o^*} : w \mapsto M_{\delta^*(q_o,w)}$;

system behavior $f\colon X_o^* \longrightarrow Y : w \mapsto \beta[\delta^*(q_o,w)]$.

This does indeed match the 'classical' formulations of sequential machine theory.

As we noted in Section 3 of [1], it has become standard category theory to generalize the notion of one-to-one maps and onto maps between sets to obtain classes $\mathcal{E}$ and $\mathcal{M}$ (standing for generalized epi and mono respectively) of morphisms in a category $\mathcal{K}$. Although in any category there are epi maps and mono maps, the maps in $\mathcal{E}$ or $\mathcal{M}$ may form proper subcategories of these classes of maps. Just as in *Set*, $f: A \to B$ always can be thought of as the composition of an onto map followed by a one-to-one map, so one can postulate that in an arbitrary category $\mathcal{K}$, any morphism f has $f = gh$ for $h \in \mathcal{E}$, $g \in \mathcal{M}$. With several additional postulates, $\mathcal{K}$ is said to have an image factorization system.

*1.19* *DEFINITION*: An X-system $(X,Q,\delta,I,\tau,Y,\beta)$ is *reachable* if X is input and $r: IX^{@} \to Q$ is in $\mathcal{E}$ ("all states can be reached by proper choice of inputs"). It is *observable* if X is output and $\sigma: Q \to YX_{@}$ is in $\mathcal{M}$ ("different states can, in theory, be distinguished by input tests").

Sometimes a class $\mathcal{E}$ of epimorphisms is defined without an $\mathcal{M}$; in this case we can speak of "reachable" without an observability theory, but the crucial definition of "minimal realization" (see *1.21* below) is still possible.

Generalizing Definition *4.1* of [2] we have:

*1.20* *DEFINITION*: Fixing I and Y, but letting the state-space Q vary, consider the category whose objects are systems $M = (X,Q,\delta,I,\tau,Y,\beta)$; and whose morphisms are *simulations* $\psi: M \to M'$ (we say M *simulates* M'), i.e. dynamorphisms $\psi: (Q,\delta) \to (Q',\delta')$ which commute with the input and output:

$$\begin{array}{ccccc} & \overset{\tau}{\nearrow} & Q & \overset{\beta}{\searrow} & \\ I & & \downarrow{\scriptstyle\psi} & & Y \\ & \underset{\tau'}{\searrow} & Q' & \underset{\beta'}{\nearrow} & \end{array} \qquad (13)$$

It is an immediate consequence of the definitions that a simulation commutes with (if X is input) the reachability and (if X is output) observability maps, i.e. if X is state-behavior, the dynamorphism $\psi$ is a simulation and satisfies the diagram

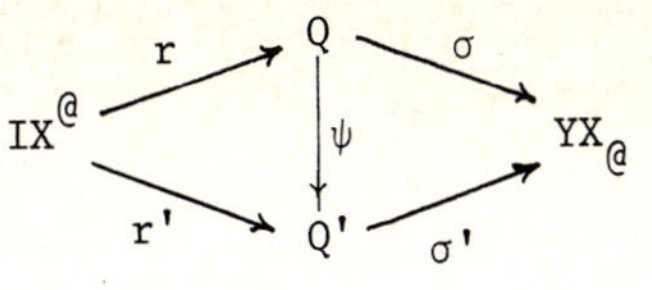


(14)

Of course, if the dynamorphism $\psi$ satisfies this diagram, it is certainly a simulation. In particular, then, *the existence of a simulation guarantees that the two machines have the same (total) response*: $f^{\blacktriangle} = \sigma r = \sigma' \psi r = \sigma' r' = f'^{\blacktriangle}$.

*1.21 DEFINITION*: Let I, Y be fixed, let $X: \mathcal{K} \to \mathcal{K}$ be an input (state-behavior) process and let $f: IX^{@} \to Y$ be an arbitrary $\mathcal{K}$-morphism (with dynamorphic coextension $f^{\blacktriangle} = f_{\#} : IX^{@} \to YX_{@}$). We say a system M is a *realization* of the (total) response f ($f^{\blacktriangle}$) if f ($f^{\blacktriangle}$) is the (total) response of M. We say M is a *minimal realization* if it is a reachable realization of f with the property that, given any other reachable realization M' of f, there exists a unique simulation $\psi: M' \to M$. In other words, M is a terminal object in the category whose objects are reachable realizations of f, and whose morphisms are simulations (composition and identities being at the level of $\mathcal{K}$). Minimal means "all states are used" and "states with the same effect in any simulation (M') are merged ($\psi$) to a single state (of M)". Minimal usually coincides with "reachable and observable" if X is state-behavior [3].

In the case of decomposable machines and sequential machines, "minimal" certainly coincides with the standard meaning.

## ADJOINT PROCESSES

In this subsection we introduce an important class of state-behavior processes--the adjoint processes. This will require the presentation of further (standard) category theory results (1.22 through 1.26) which we include here for completeness.

**1.22 DEFINITION:** We say that a collection $(in_i\colon A_i \to A \mid i \in J)$ of $\mathcal{K}$-morphisms with common codomain A is a coproduct for the collection $(A_i \mid i \in J)$ of objects of $\mathcal{K}$ if it has the property that, given any other collection $(f_i\colon A_i \to A' \mid i \in J)$, there exists a unique morphism $f\colon A \to A'$ such that

$$\begin{array}{ccc} A_i & \xrightarrow{in_i} & A \\ & \searrow^{f_i} & \downarrow f \\ & & A' \end{array} \qquad (15)$$

commutes. [All coproducts of a given family $A_i$ are isomorphic. If the index set is the set $\underline{N}$ of nonnegative integers, and the $A_i$ are equal, then their coproduct is just the countable copower, familiar from [2].]

We say $\mathcal{K}$ has finite coproducts if each finite collection $(A_1,\ldots,A_n)$ has a coproduct, the object of which we may denote $\coprod_{1\le i\le n} A_i$, or $A_1+\ldots+A_n$. Similarly, $\mathcal{K}$ has countable coproducts if every countable collection $(A_1,A_2,\ldots,A_n,\ldots)$ has a coproduct.

We say a functor X preserves coproducts if, whenever $(in_i\colon A_i \to A \mid i \in J)$ is a coproduct of the $A_i$, then $(in_iX\colon A_iX \to AX \mid i \in J)$ is a coproduct of the $A_iX$.

Just as arrow reversal turns copowers into powers [2], so may we reverse the arrows in 1.22 to define products $(\pi_i\colon A \to A_i \mid i \in J)$ by

a diagram

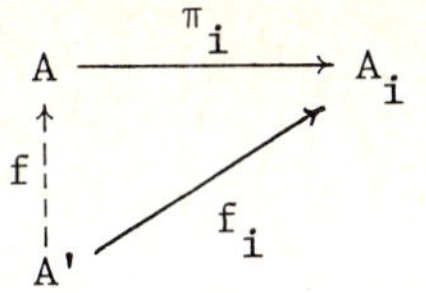


(16)

<u>1.23</u> <u>DEFINITION</u>: A functor $F: \mathcal{A} \to \mathcal{B}$ has a <u>right adjoint</u> if there exists a functor $F^{\bullet}: \mathcal{B} \to \mathcal{A}$ (the right adjoint of F) such that to each B in $\mathcal{B}$ there corresponds a $\mathcal{B}$-morphism $BFF^{\bullet} \xrightarrow{\varepsilon} B$ such that to each $\mathcal{B}$-morphism $g: AF \to B$ there corresponds a unique $\mathcal{A}$-morphism $\phi: A \to BF^{\bullet}$ such that

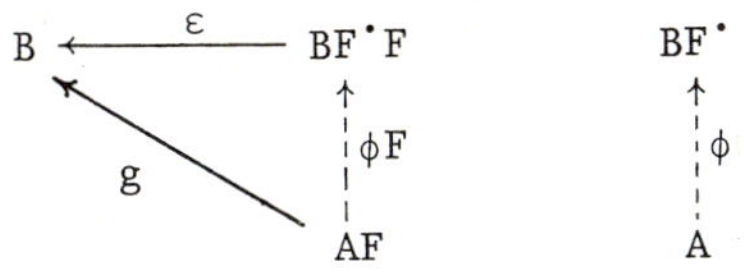


(17)

commutes.

Notice that given a $\phi: A \to BF^{\bullet}$, one immediately has a $g: AF \to B$ by $g = \varepsilon \cdot \phi F$. Indeed, adjunctions can be thought of as a bijection between sets of morphisms:

$$\frac{AF \xrightarrow{g} B}{A \xrightarrow{\phi} BF^{\bullet}} .$$

The reader for whom this concept is new could profitably see how <u>1.13</u> specializes this definition.

<u>1.24</u> <u>EXAMPLE</u>: Let <u>Vect</u> = <Vector Spaces and Linear Maps>, and let $F$: <u>Vect</u> $\to$ <u>Vect</u> be the identity functor $(Q \mapsto Q;\ f \mapsto f)$. Then F is its own right adjoint--setting $F = F^{\bullet} = id$ , $\varepsilon = id_B$, we have that (17) is satisfied with $\phi = g$ :

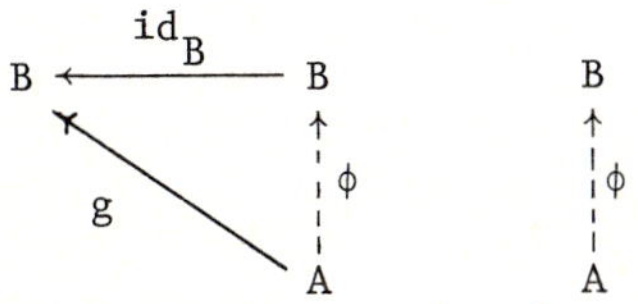

1.25 EXAMPLE: Let Set = <Sets and Maps>, and let $X = -\times X_o$ : Set $\to$ Set be the functor $Q \mapsto Q\times X_o$; $f \mapsto f\times X_o$ where

$$f\times X_o: Q\times X_o \to Q'\times X_o : (q,x) \mapsto (f(q),x) .$$

Then $X = -\times X_o$ has right adjoint $(-)^{X_o}$ which sends Q to the set $Q^{X_o}$ of all maps from $X_o$ to Q. $\varepsilon: B^{X_o}\times X_o \to B$ is the evaluation $(f,x) \mapsto f(x)$, and we have that (17) is satisfied on taking $\phi(a): X_o \to B : x \mapsto g(a,x)$ :

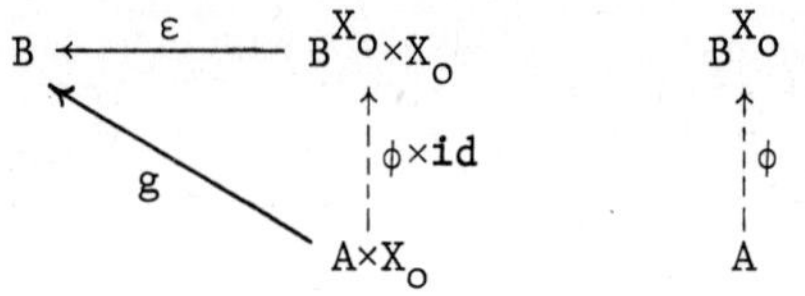


The following is a standard result connecting the ideas of coproduct and right adjoint:

1.26 LEMMA: If $F: \mathcal{A} \to \mathcal{B}$ has a right adjoint, then F preserves coproducts.

Proof: Let $(A_i \xrightarrow{in_i} A \mid i \in I)$ be a coproduct. To show that $(A_iF \xrightarrow{in_iF} AF \mid i \in I)$ is also a coproduct, we must show how to uniquely complete the diagram

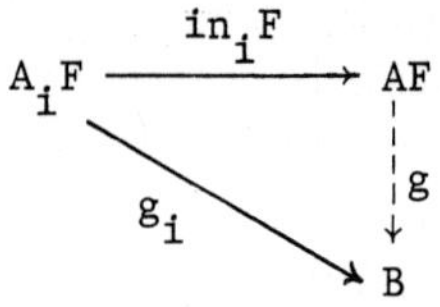


(18)

when $g_i: A_iF \to B$, except for the specification of domain and codomain, is an arbitrary collection of $\mathcal{B}$-morphisms. Let us define $\phi_i$, for each $i \in I$, by the following variant of (17):

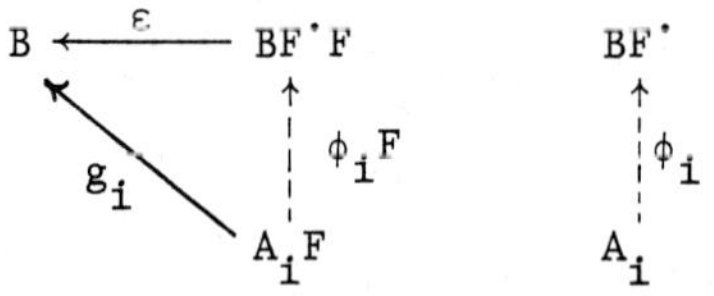

We may then use the fact that $(A_i \xrightarrow{in_i} A)$ is a coproduct to define $\phi: A \to BF^{\bullet}$ by the rule

$$\begin{array}{ccc} A_i & \xrightarrow{in_i} & A \\ & \searrow & \downarrow \phi \\ & & BF^{\bullet} \end{array}$$

Now the $g: AF \to B$ corresponding to $\phi: A \to BF^{\bullet}$ in the style of (17) is just $g = \varepsilon \cdot \phi F$, and it is a straightforward exercise to check that this $g$ is the unique solution of (18). □

<u>1.27</u> <u>CONSTRUCTION</u>: In preparation for delineating a class of input processes, we note the following construction of a functor $X^*$ from a functor $X: \mathcal{K} \to \mathcal{K}$.

Let $(\mathcal{A} \xrightarrow{H_\alpha} \mathcal{B} \mid \alpha \in I)$ be a collection of functors; and assume that $\mathcal{B}$ has I-indexed coproducts. Then we can form the functor

$$H = \coprod_{\alpha \in I} H_\alpha$$

(uniquely up to isomorphism) defined on objects by

$$AH = \coprod_{\alpha} AH_\alpha$$

and on morphisms by

$$\begin{array}{ccccc} A & & \coprod_\alpha AH_\alpha & \xleftarrow{in_\beta} & AH_\beta \\ \downarrow f & \longmapsto & \downarrow fH & & \downarrow fH_\beta \\ A' & & \coprod_\alpha A'H_\alpha & \xleftarrow{in_\beta} & BH_\beta \end{array}$$

[Identify $f_i: A_i \to A'$ in (15) with $in_\beta \cdot fH_\beta: AH_\beta \to \coprod_\alpha A'H_\alpha$ here.]
It is then straightforward to check that $\coprod_\alpha H_\alpha$ is a functor by considering its action on identities and composition of morphisms.

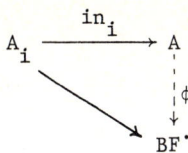

Now, given any functor $X\colon \mathcal{K} \to \mathcal{K}$, define $X^n$ for any $n \geq 0$ by the rules

$$X^0 = \mathrm{id}$$

$$X^{n+1} = X^n \cdot X \qquad \text{for } n \geq 0$$

and then set

$$X^* = \coprod_{n \geq 0} X^n .$$

And now we have a statement on the existence of input processes:

1.28 THEOREM: Let $\mathcal{K}$ have, and let $X\colon \mathcal{K} \to \mathcal{K}$ preserve, countable coproducts. Then X is an input process and $X^@ = X^*$.

Proof: If $IX^@$ is to be $IX^*$ we must define the morphisms $\mu_0\colon IX^*X \to IX^*$ and $\eta\colon I \to IX^@$ described in 1.11. Since X preserves the coproduct $IX^n \xrightarrow{in_n} IX^*$, we have that

$$IX^{n+1} = IX^n X \xrightarrow{in_n X} IX^* X$$

is also a coproduct. Thus we may define $\mu_0$ by the obvious rule

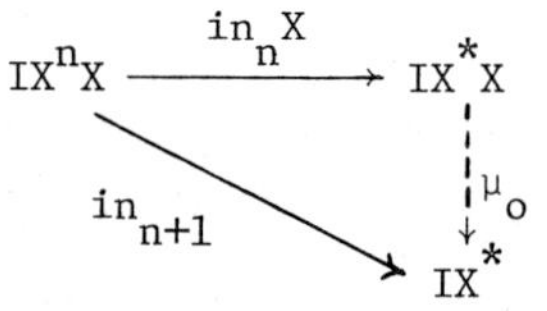


We define $\eta\colon I \to IX^*$ to be simply $in_0$. Let us check that this works, i.e. that the diagrams

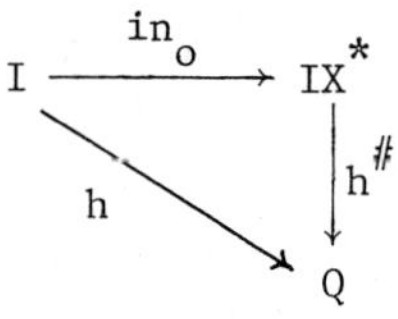


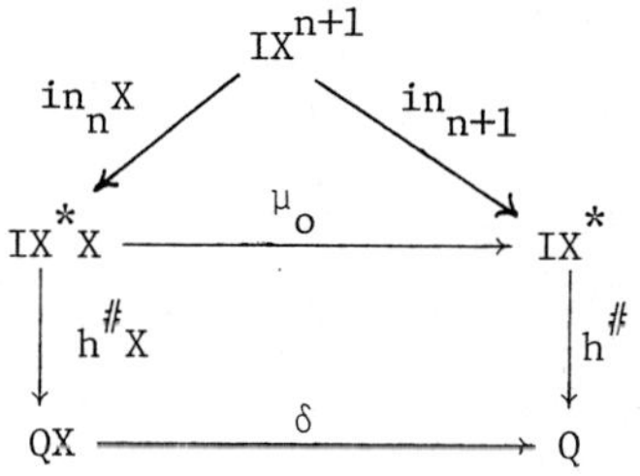


define a unique $h^{\#}\colon IX^* \to Q$. But the left-hand diagram says

$$h^{\#} \cdot in_0 = h$$

while the right-hand diagram asserts that

$$h^{\#}\cdot in_{n+1} = \delta\cdot h^{\#}X\cdot in_n X = \delta\cdot(h^{\#}\cdot in_n)X \qquad n \geq 0 .$$

Evidently, $h^{\#}\cdot in_o$, $h^{\#}\cdot in_1$, ... can be defined successively, so that these equations define the unique $h^{\#}$ which satisfies the diagrams. □

A special case of this result (where X is of the form $-\otimes X_o$ for a suitable "tensor product" $\otimes$ and object $X_o$) lies at the heart of the theories of Goguen [13] and Ehrig-Kreowski [14].

Using 1.26, we have

1.29 COROLLARY: If $\mathcal{K}$ has countable coproducts and $X: \mathcal{K} \to \mathcal{K}$ has a right adjoint, then X is an input process, and $X^{@} = X^{*}$. □

The functor $X = -\times X_o$: **Set** $\to$ **Set** is certainly included within the ambit of the above discussions. We could have obtained the definitions of $IX^{@}$ and so on in the early part of 1.16 once one checks preservation of coproduct, or reaches the observation of 1.25 that the right adjoint of $-\times X_o$ is $(-)^{X_o}$.

The question arises of when X is also an output process, i.e. when X is state behavior. We have the following result from [3]:

1.30 THEOREM: Let $\mathcal{K}$ have countable coproducts and products, and let X have a right adjoint. Then X is state-behavior.

Proof: In outline, this can be argued by observing that the right adjoint $X^{\bullet}$ of X defines a functor $\mathcal{K}^{op} \to \mathcal{K}^{op}$ which itself has a right adjoint. It then is an input process. Taking duals leads to the conclusion that X is an output process. One has the important result also that

$$X_{@} = \prod_{n\geq 0} (X^{\bullet})^n . \qquad (19)$$ □

For convenience, and because of the importance of the concept, we formalize this class of processes:

1.31 DEFINITION: Let $X: \mathcal{K} \to \mathcal{K}$ be a functor. Then X is an adjoint process if $\mathcal{K}$ has countable coproducts and products, and X has a right adjoint.*

1.32 EXAMPLES:

(i) In case $X = -\times X_o : \underline{Set} \to \underline{Set}$, we have X is an adjoint process. As computed earlier, $IX^@ = I\times X_o^* \cong \coprod_{n\geq 0} IX^n$ and $YX_@ = Y^{X_o^*} = \prod_n Y^{X_o^n}$.

(ii) In case $X = id_{\mathcal{K}}$ and $\mathcal{K}$ has countable coproducts and products, X is an adjoint process. As computed earlier, $IX^@ = I^§$ and $YX_@ = Y_§$.

(iii) Let X be as in the time-varying linear system example of 1.8. Then X again is adjoint. For products are formed t-wise in Vect, and coproducts are "weak direct sums". Define $(Q_t)X^\cdot = (Q_{t+1})$ and, for $(f_t): (Q_t) \to (R_t)$, $(f_t)X^\cdot = (\bar{f}_t) : (Q_{t+1}) \to (R_{t+1})$. Then $X^\cdot$ can be shown to be a functor and the right adjoint of X, using 1.23.

As discussed in [3], adjoint processes are rich enough to include the processes appearing in sequential machines, nondeterministic machines, Boolean machines, metric machines and topological machines.

---

* Actually, if X is state-behavior, has a right adjoint and $\mathcal{K}$ is simply assumed to have coproducts, a nontrivial argument shows that $YX_@$ has to be $\prod_{n\geq 0} Y(X^\cdot)^n$. So in this case, the specific assumption that $\mathcal{K}$ has products is not necessary.

MORE ON DYNAMORPHISMS

In this final subsection of Section 1, we present some useful properties of dynamorphisms. We start by examining approximants to dynamorphic extensions:

1.33 LEMMA: Given any morphism $a\colon A \to Q$, and a dynamics $\delta\colon QX \to Q$, let $a^{\#}\colon AX^{@} \to Q$ be the dynamorphic extension of $a$

$$\begin{array}{ccc} A & \xrightarrow{\eta} & AX^{@} \\ & \searrow^{a} & \downarrow{a^{\#}} \\ & & Q \end{array} \qquad\qquad \begin{array}{ccc} (AX^{@})X & \xrightarrow{\mu_o} & AX^{@} \\ {a^{\#}X}\downarrow & & \downarrow{a^{\#}} \\ QX & \xrightarrow{\delta} & Q \end{array}$$

and let $a_n$ be the n-step approximant to $a^{\#}$

$$a_n = AX^n \xrightarrow{aX^n} QX^n \xrightarrow{\delta^{(n)}} Q$$

where $\delta^{(n)}$ is the n-fold iteration of the dynamics $\delta$ defined inductively by

$$\delta^{(0)} = id_Q\colon Q \to Q \quad \text{while} \quad \delta^{(n+1)} = QX^{n+1} \xrightarrow{\delta\cdot X^n} QX^n \xrightarrow{\delta^{(n)}} Q.$$

By an easy induction, we have also $\delta^{(n+1)} = QX^{n+1} \xrightarrow{\delta^{(n)}X} QX \xrightarrow{\delta} Q$.

Clearly, then, $a_{n+1} = \delta\cdot a_n X$. As a special case, let $\eta_n$ be the n-step inclusion

$$\eta_n\colon AX^n \xrightarrow{\eta X^n} AX^{@}X^n \xrightarrow{\mu_o^{(n)}} AX^{@}$$

where $\mu_o^{(n)}$ is defined from $\mu_o$ just as $\delta^{(n)}$ was defined from $\delta$. Again, $\eta_{n+1} = \mu_o\cdot\eta_n X$. Then for all $n \geq 0$ we have the commutativity of

$$\begin{array}{ccc} AX^n & \xrightarrow{\eta_n} & AX^{@} \\ & \searrow^{a_n} & \downarrow{a^{\#}} \\ & & Q \end{array} \qquad (20)$$

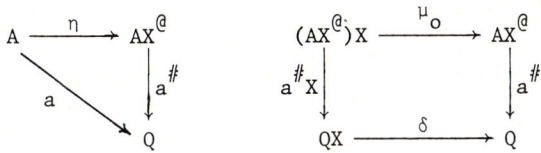

Proof: The case $n = 0$ is just $a = a^{\#}\cdot\eta$. For the induction step, we assume the commutativity of (20) for a given n, and show it also holds when n is replaced by n+1. We have:

$$
\begin{array}{ccccc}
AX^{n+1} & \xrightarrow{\eta_n X} & AX^{@}X & \xrightarrow{\mu_o} & AX^{@} \\
 & \searrow^{a_n X} & \downarrow{\scriptstyle a^{\#}X} & & \downarrow{\scriptstyle a^{\#}} \\
 & & QX & \xrightarrow{\delta} & Q
\end{array}
$$

The triangle is obtained from (20) and the fact that X is a functor; while the square says that $a^{\#}$ is a dynamorphism. Thus

$$a_{n+1} = \delta \cdot a_n X = a^{\#} \cdot \mu_o \cdot \eta_n X = a^{\#} \cdot \eta_{n+1} . \qquad \square$$

In case X is an adjoint process, we can show that

$$\eta_n : AX^n \longrightarrow AX^{@} = in_n : AX^n \longrightarrow \coprod_{n \geq 0} AX^n .$$

This is easy to see by an inductive argument. As a basis, observe that $\eta_1 = \mu_o \cdot \eta X = \mu_o \cdot in_o X = in_1$, using the last diagram in 1.28. Suppose that $\eta_n = in_n$. Then $\eta_{n+1} = \mu_o \cdot \eta_n X = \mu_o \cdot in_n X = in_{n+1}$, again using 1.28.

In the next lemma, we shall label $\mu_o$ and $\eta$ with the object A which generates the free dynamics with which they are associated, so that we have $A\eta : A \longrightarrow AX^{@}$; $A\mu_o : AX^{@}X \longrightarrow AX^{@}$. Just as we may consider a pair $(q,x) \in Q \times X_o$ for a sequential machine as a pair $(q,(x)) \in Q \times X_o^*$ by identifying the input x with the input string (x); so may we, given any input process $X : \mathcal{K} \longrightarrow \mathcal{K}$ and any object A of $\mathcal{K}$, define (by 1.33)

$$A\eta_1 = AX \xrightarrow{(A\eta)X} AX^{@}X \xrightarrow{A\mu_o} AX^{@} .$$

Again, given any morphism $f : A \longrightarrow B$ we can define a morphism $fX^{@} : AX^{@} \longrightarrow BX^{@}$ by the diagram

$$
\begin{array}{ccc}
A & \xrightarrow{A\eta} & AX^{@} \\
{\scriptstyle f}\downarrow & & \downarrow{\scriptstyle fX^{@} = (B\eta \cdot f)^{\#}} \\
B & \xrightarrow{B\eta} & BX^{@}
\end{array}
\qquad (21)
$$

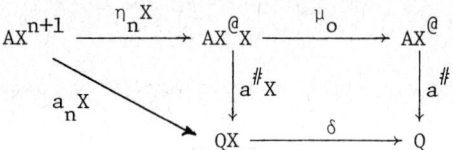

[It is a straightforward exercise to show that, with this action on morphisms, $X^@$ becomes a functor $\mathcal{K} \to \mathcal{K}$.] Then we have the following lemma:

1.34 LEMMA: For all $f: A \to B$, we have the commutativity[†] of

$$\begin{array}{ccc} AX & \xrightarrow{A\eta_1} & AX^@ \\ {\scriptstyle fX}\downarrow & & \downarrow{\scriptstyle fX^@} \\ BX & \xrightarrow{B\eta_1} & BX^@ \end{array} \qquad (22)$$

Proof: We simply use the definition of $A\eta_1$ and $B\eta_1$ to expand (22) to the diagram

$$\begin{array}{ccccc} AX & \xrightarrow{(A\eta)X} & AX^@X & \xrightarrow{A\mu_o} & AX^@ \\ {\scriptstyle fX}\downarrow & & \downarrow{\scriptstyle fX^@X} & & \downarrow{\scriptstyle fX^@} \\ BX & \xrightarrow{(B\eta)X} & BX^@X & \xrightarrow{B\mu_o} & BX^@ \end{array}$$

and note that the left-hand square commutes by the fact that (21) commutes and that X is a functor; while the right-hand square just says that $fX^@ = (B\eta \cdot f)^{\#}$ is a dynamorphism. □

1.35 DEFINITION: The run map $\delta^@: QX^@ \to Q$ of a dynamics $(Q,\delta)$ is the dynamorphic extension

$$\begin{array}{ccc} Q & \xrightarrow{Q\eta} & QX^@ \\ & {\scriptstyle id_Q}\searrow & \downarrow{\scriptstyle \delta^@ = (id_Q)^{\#}} \\ & & Q \end{array}$$

Lemma 1.33 then immediately yields:

1.36 COROLLARY: $\delta^@ \cdot Q\eta_1 = (id_Q)_1 = \delta$ . □

---

† The category theorist will recognize that (21) and (22) say that $\eta$ and $\eta_1$ are natural transformations, but this general notion need not detain us here.

Finally we observe that "a dynamorphism, since it commutes with one-step transitions, must commute with all transitions":

**1.37 LEMMA**: Given any dynamorphism $\psi: (Q,\delta) \rightarrow (Q',\delta')$, we have the commutativity of

$$\begin{array}{ccc} QX^@ & \xrightarrow{\delta^@} & Q \\ {\scriptstyle \psi X^@}\downarrow & & \downarrow{\scriptstyle \psi} \\ Q'X^@ & \xrightarrow{(\delta')^@} & Q' \end{array}$$

Proof: $\psi\cdot\delta^@\cdot Q\eta = \psi\cdot id_Q = \psi$ since $\delta^@ = (id_Q)^\#$.

$(\delta')^@\cdot\psi X^@\cdot Q\eta = (\delta')^@\cdot Q'\eta\cdot\psi$ by (21) with $f = \psi: Q \rightarrow Q'$

$= id_{Q'}\cdot\psi = \psi$ since $(\delta')^@ = (id_{Q'})^\#$

Thus $\psi\cdot\delta^@ = \psi^\# = (\delta')^@\cdot\psi X^@$, and so our diagram commutes. □

## 2. Nerode Equivalence Approach

In Section 3 of [2] we presented the notion of an image factorization system; and then showed in Section 4 how this yielded a simple minimal realization theory for decomposable systems. In [3] we extended this approach to X-systems for any state-behavior process X (yielding a theory akin to that of Bainbridge [21]).

However, in [1], we built upon the Nerode equivalence approach to minimal realization to yield a theory applicable to input processes, even if they are not state-behavior. Our aim in this section is to give an internal Nerode approach which is both more elegant and more powerful than the internal approach of [1].

In what follows, let $X\colon \mathcal{K} \to \mathcal{K}$ be a fixed input process, I, Y fixed objects of $\mathcal{K}$.

2.1 DEFINITION: Given a response morphism $f\colon IX^@ \to Y$, we say that a pair of morphisms $\alpha,\gamma\colon E \to IX^@$ are abstractly f-equivalent if $f\alpha^\# = f\gamma^\#$ where $\alpha^\#$ (resp. $\gamma^\#$) is the unique dynamorphic extension $EX^@ \to IX^@$ of $\alpha$ (resp. $\gamma$).

We then say that $\alpha,\gamma\colon E_f \to IX^@$ is the abstract Nerode equivalence of f if $f\alpha^\# = f\gamma^\#$ holds and if, wherever $f(\alpha')^\# = f(\gamma')^\#$ also holds for $\alpha',\gamma'\colon E' \to IX^@$ there exists a unique $\psi\colon E' \to E_f$ such that

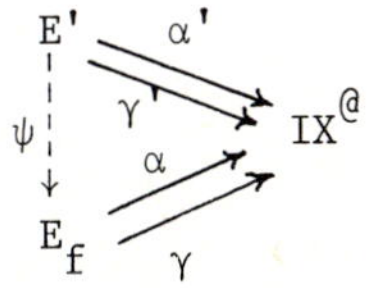


(1)

commutes, i.e. $\alpha\psi = \alpha'$ and $\gamma\psi = \gamma'$.

When $X = -\times X_o$: *Set* $\rightarrow$ *Set*, this reduces [1] to the familiar condition that two strings $w_1$ and $w_2$ of $X_o^*$ are equivalent iff $f(w_1 w) = f(w_2 w)$ for all $w$ in $X_o^*$.

It is clear that the abstract Nerode equivalence is a terminal object in a category (recall [2, *2.6*]) with morphisms of the kind depicted in (1), and so is unique up to isomorphism (by Lemma *2.8* of [2]).

For our work on tree automata, a minor variant on this definition will be required. A pair $t,u\colon A \rightarrow B$ is termed *reflexive* if there exists a $v\colon B \rightarrow A$ making the following diagram commutative:

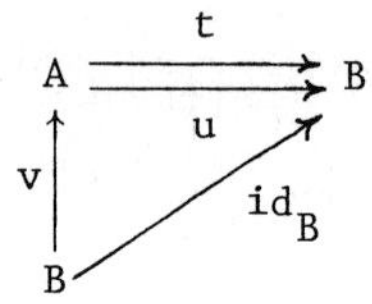


In *Set*, this simply means that every $b \in B$ is such that for some $a \in A$, $t(a) = u(a) = b$. The abstract Nerode equivalence of $t$, when it exists, is a reflexive pair, because by the Nerode property, there exists $\psi$ making the following diagram commutative:

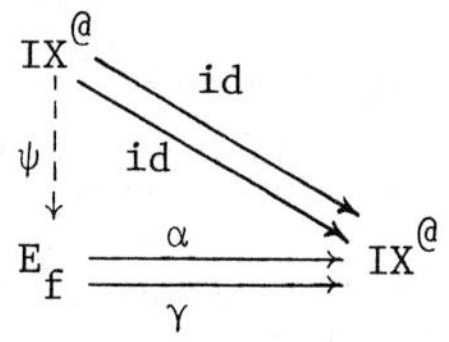


We say that $\alpha,\gamma\colon E_f \underset{\gamma}{\overset{\alpha}{\rightrightarrows}} IX^@$ is the *reflexive Nerode equivalence of f* if $f\alpha^\# = f\gamma^\#$ and if, whenever $f(\alpha')^\# = f(\gamma')^\#$ for a *reflexive* $\alpha',\gamma'\colon E' \rightarrow IX^@$ there exists a unique $\psi$ making (1) commutative.

Evidently, any abstract Nerode equivalence is a reflexive Nerode equivalence, but not necessarily conversely. (Indeed, we shall exhibit in Section 3 a reflexive Nerode equivalence which is not an abstract Nerode equivalence.)

An external version [1] of 2.1 results by iterating X:

**2.2 DEFINITION:** Given a response $f\colon IX^{@} \longrightarrow Y$, a pair of morphisms $\alpha,\gamma\colon E \longrightarrow IX^{@}$ are *externally f-equivalent* if

$$f\cdot\mu_o^{(n)}\cdot\alpha X^n = f\cdot\mu_o^{(n)}\cdot\gamma X^n \qquad (\text{all } n \in \mathbb{N}) \tag{2}$$

where (cf. 1.33) $\mu_o^{(n)}\colon IX^{@}X^n \longrightarrow IX^{@}$ is the n-fold iteration of the free dynamics $\mu_o$. Note that 1.33 also shows that (2) can be rephrased as $f\cdot\alpha_n = f\cdot\gamma_n$, where $\alpha_n$ is the n-step approximant to $\alpha^{\#}$. $\alpha,\gamma\colon E_f \longrightarrow IX^{@}$ is the *external Nerode equivalence of* f if $\alpha,\gamma$ are externally f-equivalent and if whenever $\alpha',\gamma'\colon E' \longrightarrow IX^{@}$ are externally f-equivalent there exists a unique $\psi\colon E' \longrightarrow E_f$ as in (1).

The Nerode equivalence approach to minimal realization is: "the state-space $Q_f$ of the minimal realization is defined by $Q_f = IX^{@}/E_f$". The categorical problem is one of capturing "how to divide out by the equivalence relation $E_f$". This is what coequalizers are for:

**2.3 DEFINITION:** We say a $\mathcal{K}$-morphism $A \xrightarrow{h} B$ is a *coequalizer* iff there exists a pair $p_1,p_2\colon R \longrightarrow A$ of morphisms such that $h\cdot p_1 = h\cdot p_2$, and such that whenever $A \xrightarrow{h'} B'$ satisfies $h'\cdot p_1 = h'\cdot p_2$, there is a unique $\mathcal{K}$-morphism $B \xrightarrow{\psi} B'$ such that $\psi\cdot h = h'$

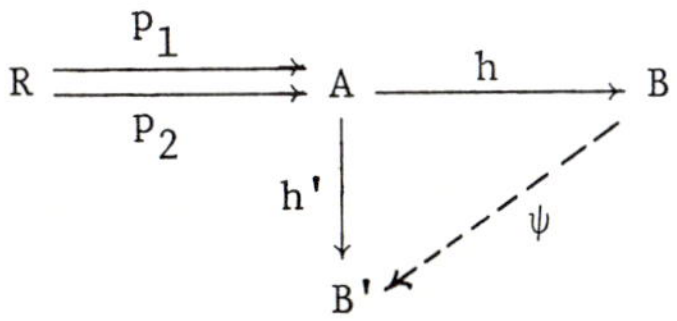


(3)

In this situation, we call h the *coequalizer of* $p_1$ *and* $p_2$, and write $h = \mathrm{coeq}(p_1,p_2)$. Standard category theory arguments establish the uniqueness up to isomorphism of $\mathrm{coeq}(p_1,p_2)$ if it exists.

An immediate consequence of the definition is the following standard property.

2.4 LEMMA: Every coequalizer is an epimorphism.

Proof: Suppose $h = \mathrm{coeq}(p_1,p_2)$ and that $k_1 \cdot h = k_2 \cdot h$ for some $k_1$ and $k_2$. We must prove that $k_1 = k_2$. But if we take $h' = k_1 \cdot h$ in 2.3 -- which we may since $h' \cdot p_1 = k_1 \cdot (h \cdot p_1) = k_1 \cdot (h \cdot p_2) = h' \cdot p_2$ -- we see that there is a unique $\psi$ such that $\psi \cdot h = h'$. But $h' = k_1 \cdot h = k_2 \cdot h$, by hypothesis, and so we must have that $k_1 = k_2$. Thus h is an epimorphism. □

2.5 EXAMPLE: In Set, $\mathrm{coeq}(p_1,p_2)$ can be found in the following way. Say that two elements $a_1, a_2 \in A$ are strictly equivalent if there exists some $r \in R$ for which $p_1(r) = a_1$, $p_2(r) = a_2$, and say that two elements $a_1, a_2 \in A$ are equivalent if $a_1 = a_2$ or if there exists for some $b_0, b_1, \ldots, b_n \in A$, adjacent pairs of the sequence $(a_1, b_0, b_1, \ldots, b_n, a_2)$ which are strictly equivalent. This defines an equivalence relation $\bar{R}$, which is generated by the image of R, $\{(p_1(r), p_2(r)) \mid r \in R\}$, in $A \times A$. Define h as the canonical projection $A \longrightarrow A/\bar{R}$.

Suppose that $h' \cdot p_1 = h' \cdot p_2$. For $\alpha \in A/\bar{R}$, define $\psi(\alpha) = h'(a_1)$ where $a_1$ is any member of the equivalence class $A/\bar{R}$. If $a_1, a_2$ are in the same equivalence class, set up the sequence $(a_1, b_0, b_1, \ldots, b_n, a_2)$ with adjacent pairs strictly equivalent. Then $h'(a_1) = hp_1(r) = hp_2(r) = h'(b_0)$ $= \ldots = h'(a_2)$. This shows that $\psi$ is well defined; $\psi h = h'$ is immediate, and $\psi$ is obviously unique.

In Vect, $\mathrm{coeq}(p_1,p_2)$ exists and is obtained as $A/\ker(p_1 - p_2)$.

In both Set and Vect, coequalizers coincide with the epi maps.

The Nerode equivalence theorem of [1] generalizes easily into the framework of Definition 2.1. The four postulates have been well motivated in [1]. We shall see below that they are satisfied in a wide range of circumstances:

2.6 THEOREM: Let the response morphism $f\colon IX^{@} \to Y$ satisfy the following four postulates*:

Postulate 1: f has an abstract Nerode equivalence $E_f \underset{\gamma}{\overset{\alpha}{\rightrightarrows}} IX^{@}$.

Postulate 2: $r_f = \mathrm{coeq}(\alpha,\gamma)\colon IX^{@} \to Q_f$ exists.

Postulate 3: There exists a dynamics $(Q_f,\delta_f)$ with $r_f\colon (IX^{@},\mu_o) \to (Q_f,\delta_f)$ a dynamorphism.

Postulate 4: X is such that if an X-dynamics $(Q,\delta)$ has reachability map $r\colon IX^{@} \to Q$ with r a coequalizer, then either rX or $rX^{@}$ is an epimorphism.

Then f has a minimal coequalizer-reachable (i.e., $\mathcal{E}$ is the class of coequalizers) realization.

If in Postulate 1 the assumption of abstract Nerode equivalence is replaced by reflexive Nerode equivalence, then f has a minimal reflexive-coequalizer-reachable realization (where a reflexive coequalizer is one coequalizing a reflexive pair).

Proof: By definition of $\alpha^{\#}$, we have $\alpha^{\#}\eta = \alpha$. By 2.1, we have $f\alpha^{\#} = f\gamma^{\#}$. Thus

$$f\cdot\alpha = f\cdot\alpha^{\#}\cdot\eta = f\cdot\gamma^{\#}\cdot\eta = f\cdot\gamma\ .$$

Then we may use the fact that $r_f = \mathrm{coeq}(\alpha,\gamma)$:

---

* The postulates are identical with those of [1], save that the abstract rather than external Nerode equivalence is used, and Postulate 4 is mildly loosened, by allowing $rX^{@}$, rather than rX, to be an epimorphism.

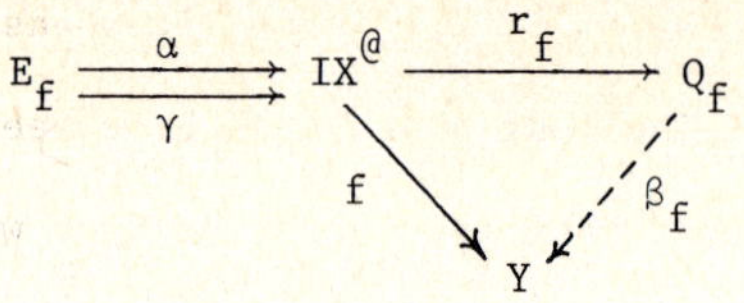


(4)

to define the unique $\beta_f$ such that $\beta_f \cdot r_f = f$.

Defining $\tau_f\colon I \to Q_f$ to be $r_f \cdot \eta$, Postulate 3 ensures that $r_f$ is the reachability map of $M_f = (X, Q_f, \delta_f, I, \tau_f, Y, \beta_f)$ which is thus a coequalizer-reachable realization of f.

To see that $M_f$ is minimal, let M be a reachable realization of f, with reachability map $r = \text{coeq}(a,b)\colon IX^@ \to Q$. We must exhibit a unique simulation $\Gamma\colon M \to M_f$.

From the diagram

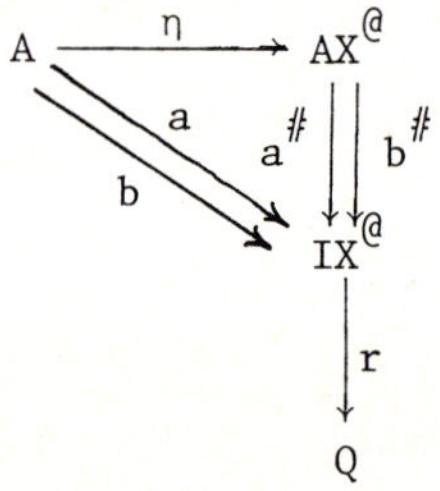


and the fact that r is a dynamorphism, we conclude that $(ra)^\# = ra^\#$, $(rb)^\# = rb^\#$. Since $r \cdot a = r \cdot b$ by the coequalizer property we have that $(ra)^\# = (rb)^\#$. Then

$$fa^\# = \beta \cdot r \cdot a^\# = \beta(ra)^\# = \beta(rb)^\# = fb^\#$$

and so by (1) there exists a unique $\psi\colon A \to E_f$ completing the diagram

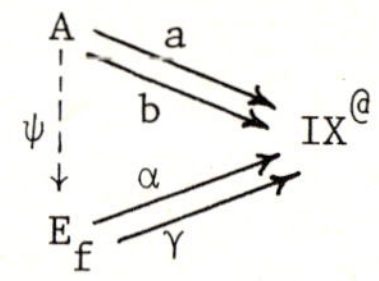


which we may then insert into

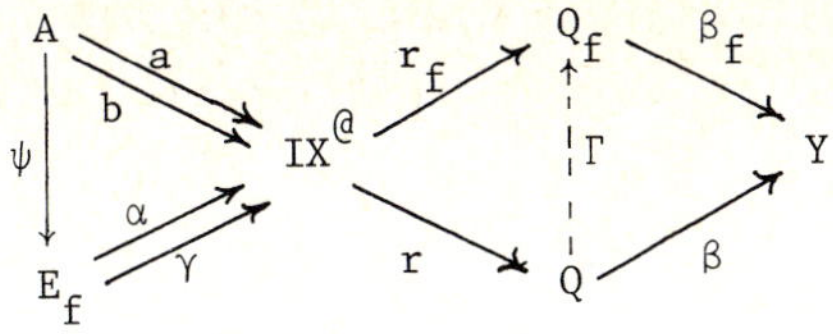


(5)

But then the fact that $r_f \cdot a = r_f \cdot \alpha \cdot \psi = r_f \cdot \gamma \cdot \psi = r_f \cdot b$ induces (by the coequalizer property of r) a unique $\Gamma$ such that $\Gamma \cdot r = r_f$. As r (being a coequalizer) is epi, $\beta \cdot r = \beta_f \cdot r_f = \beta_f \cdot \Gamma \cdot r$ implies that $\beta = \beta_f \cdot \Gamma$.

In case $E_f \overset{\alpha}{\underset{\gamma}{\rightrightarrows}} IX^@$ is a reflexive Nerode equivalence, the argument is the same, save that $a,b: A \to IX^@$ is a reflexive pair.

Finally, we use Postulate 4 to see that $\Gamma$ is a dynamorphism.

First, suppose rX is an epimorphism. Then

$$\begin{aligned} \delta_f \cdot \Gamma X \cdot rX &= \delta_f \cdot r_f X = r_f \cdot \mu_o && \text{(by Postulate 3)} \\ &= \Gamma \cdot r \cdot \mu_o \\ &= \Gamma \cdot \delta \cdot rX && \text{(again by Postulate 3)} \end{aligned}$$

Then $\delta_f \cdot \Gamma X = \Gamma \cdot \delta$ since rX is epi, and this is the desired dynamorphic property. This completes the proof of the theorem in case rX is epi.

Otherwise suppose $rX^@$ is epi. To prove that $\Gamma$ is a dynamorphism consider

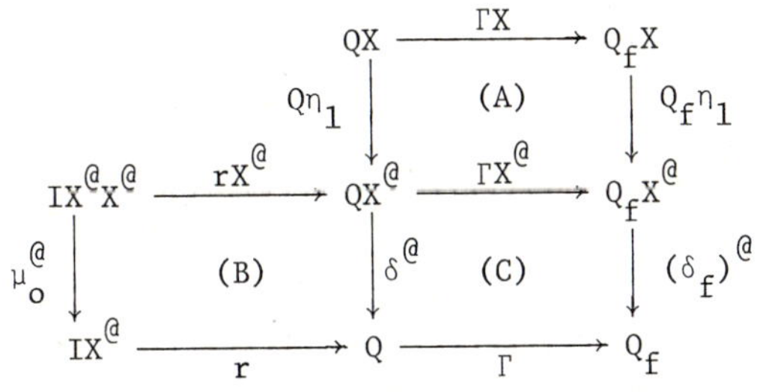


Then (A) commutes by Lemma 1.34. Since r is a dynamorphism, (B) commutes by Lemma 1.37. Similarly, since $\Gamma r = r_f$ is a dynamorphism, the perimeter (B)(C) commutes; but then--using the hypothesis that $rX^@$ is epi--(C) commutes.

By Corollary 1.36, the perimeter of (A)(C) is the statement that $\Gamma$ is a dynamorphism. □

At first sight, it might seem that verification of postulates 1 to 4 would be a major bar to application of this theorem in many cases, and the postulates look very much as if they are tailored to guarantee the desired results, rather than reflecting commonly encountered properties. However, we shall see below that the postulates can be satisfied if certain common conditions are fulfilled. Further, we shall see later that in a wide variety of cases, all coequalizers are reflexive coequalizers, so that the reflexive assumption is not a major one. Recalling Lemma 1.33 ($a^{\#} \cdot \eta_n = a_n$ for any $a: A \to Q$), we can verify that the external Nerode theory [1] is a special case of the abstract approach. [Actually, this claim is open to a charge that it is subjectively based. As the following theorem shows, it is not true that if the external Nerode equivalence holds, then the abstract equivalence holds. Rather, we need side conditions associated with the external Nerode equivalence for the abstract equivalence to hold. To the extent that these side conditions are minor, then, the abstract generalizes the external.]

2.7 THEOREM: Let $f: IX^{@} \to Y$ satisfy

Postulate 1': f has an external Nerode equivalence $\alpha, \gamma: E_f \to IX^{@}$ satisfying postulates 2 and 3 of 2.6.

Then $E_f$ is the abstract Nerode equivalence of f.

Proof: As in the proof of 2.6, we note that $(r_f \cdot \alpha)^{\#} = r_f \alpha^{\#}$. Thus

$$f \cdot \alpha^{\#} = \beta_f \cdot r_f \cdot \alpha^{\#} = \beta_f (r_f \cdot \alpha)^{\#} = \beta_f (r_f \cdot \gamma)^{\#} = f \cdot \gamma^{\#} .$$

To show $E_f$ is indeed a Nerode equivalence in the sense of 2.1, we just note that if $T \underset{u}{\overset{t}{\rightrightarrows}} IX^{@}$ satisfy $f \cdot t^{\#} = f \cdot u^{\#}$ then, by 1.33

$$f\cdot\mu_o^{(n)}\cdot tX^n = f\cdot t_n = t^{\#}f\cdot I\eta_n = u^{\#}f\cdot I\eta_n = f\cdot u_n = f\cdot\mu_o^{(n)}\cdot uX^n$$

for all $n \in \underline{\underline{N}}$, yielding the desired $\psi\colon T \to E_f$ by postulate 1'. □

Taking into account Theorem 2.6, and a corresponding theorem of [1] using the external equivalence, it is evident that if an $f\colon IX^@ \to Y$ is guaranteed to have a minimal coequalizer-reachable realization via the external theory of [1], then it is guaranteed to have such a realization via the abstract equivalence theory of 2.6. In this sense, the abstract equivalence is a true generalization of the external equivalence.

In one situation of some importance, the abstract and external Nerode equivalences become the same thing. For adjoint machines, recall from 1.33 that $\alpha_n$ and $\alpha^{\#}$ are related by

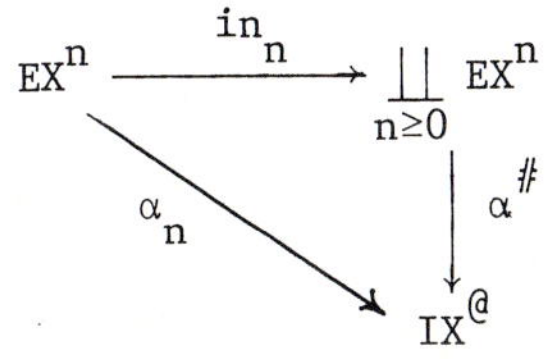


and to say that $f\alpha^{\#} = f\gamma^{\#}$ is the same thing as to say $f\cdot\alpha_n = f\cdot\gamma_n$ for all $n \in \underline{\underline{N}}$.

In 3.14, we shall give an example of an $f\colon IX^@ \to Y$ for which the reflexive Nerode equivalence exists but for which the external Nerode equivalence fails.

Our task now is to see what happens to the Nerode equivalence relation when X is a state-behavior process. The key is the commutative diagram:

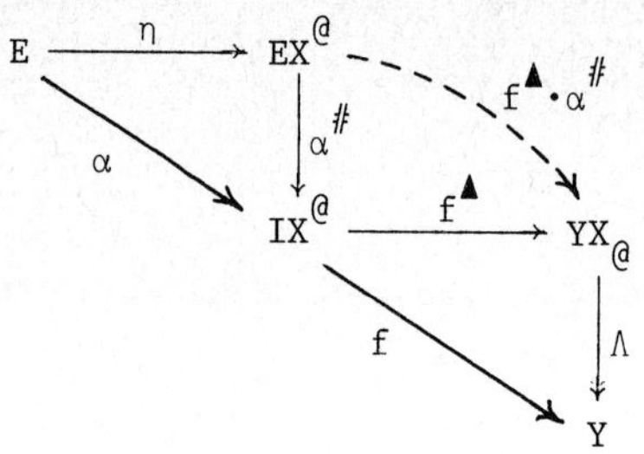


(6)

From this we read off two diagrams

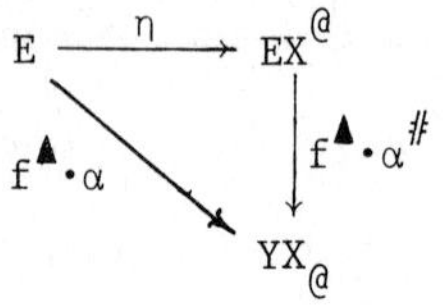


and

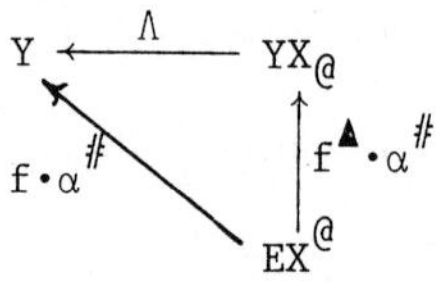


(7)

Since $f^{\blacktriangle}$ and $\alpha^{\#}$ are dynamorphisms, so is their composite, and we deduce that

$$(f^{\blacktriangle}\cdot\alpha)^{\#} = f^{\blacktriangle}\cdot\alpha^{\#} = (f\cdot\alpha^{\#})_{\#} \ .$$

Hence, it is immediate from 2.1 that:

2.8 PROPOSITION: Given $f\colon IX^{@} \to Y$ for X a state-behavior process, the pair of morphisms $\alpha,\gamma\colon E \to IX^{@}$ are abstractly f-equivalent iff

$$f^{\blacktriangle}\cdot\alpha = f^{\blacktriangle}\cdot\gamma \ . \qquad (8) \qquad \square$$

In order to find conditions under which there can be morphisms $\alpha,\gamma$ associated with a prescribed $f^{\blacktriangle}$ and satisfying (8), we recall from category theory the definition:

2.9 DEFINITION: Given a morphism $g\colon A \to B$, the pair $p,q\colon E \to A$ is called the kernel pair of g if it satisfies the equality $g\cdot p = g\cdot q$, and if, whenever the pair $p',q'\colon E' \to A$ satisfies $g\cdot p' = g\cdot q'$, there exists a unique $\psi\colon E' \to E$ with $p\cdot\psi = p'$ and $q\cdot\psi = q'$ :

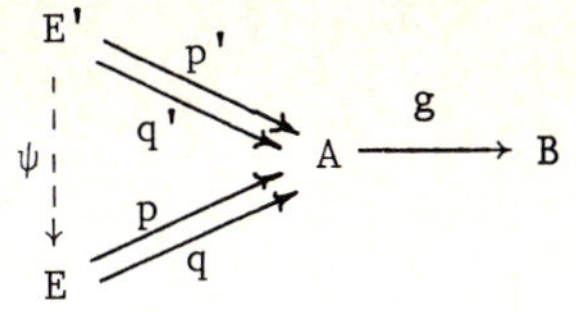


$\mathcal{K}$ has kernel pairs if every $\mathcal{K}$-morphism g has a kernel pair.

Set, Vect and most familiar categories have kernel pairs. Given $g: A \to B$, the usual construction is to set $E = \{(a_1,a_2) \mid g(a_1) = g(a_2)\} \subset A{\times}A$, with $p,q: E \to A$ the projections $p(a_1,a_2) = a_1$, $q(a_1,a_2) = a_2$.

Comparing 2.1 and 2.9 we immediately deduce from 2.8 that

2.10 PROPOSITION: Let X be a state-behavior process, and let $f^{\blacktriangle}: IX^{@} \to YX_{@}$ be a total response map. Then if $f^{\blacktriangle}$ has a kernel pair $\alpha,\gamma: E_f \to IX^{@}$, it is the abstract Nerode equivalence of f, and conversely.

□

The desired theorem asserts that the Nerode realization theorem 2.8 applies to a state-behavior process in any category equipped to "divide out by equivalence relations".

2.11 THEOREM: Let $\mathcal{K}$ be such that every morphism has a kernel pair which, in turn, has a coequalizer, and let $X: \mathcal{K} \to \mathcal{K}$ be state-behavior. Then every response morphism $f: IX^{@} \to Y$ has a coequalizer-minimal realization by the Nerode realization construction of 2.6; indeed, the minimal "state-space" object $Q_f$ is the coequalizer of the kernel pair $E_f$ of $f^{\blacktriangle}: IX^{@} \to YX_{@}$ and $E_f$ is the abstract Nerode equivalence of f.

Proof: As a result of 2.10, we see that we have only to establish postulates 3 and 4 of 2.6. The fundamental observation is that the functor $X^{@}$ has $X_{@}$ as a right adjoint:

$$\begin{array}{ll}
AX^{@} \longrightarrow B & \mathcal{K}\text{-morphism} \\
\hline
AX^{@} \longrightarrow BX_{@} & X\text{-dynamorphism} \\
\hline
A \longrightarrow BX_{@} & \mathcal{K}\text{-morphism}
\end{array}$$

Therefore, by a variant on the argument of Lemma 1.26, (see [10, p. 114, dual of Theorem 1]), $X^{@}$ preserves coequalizer diagrams, i.e. if $c = \mathrm{coeq}(a,b)$ then $cX^{@} = \mathrm{coeq}(aX^{@}, bX^{@})$. Postulate 4 is now immediate, noting 2.4. For the proof of postulate 3, argue as follows:

Let $\alpha,\gamma\colon E_f \to IX^{@}$ be the kernel pair of $f^{\blacktriangle}$, and let $r_f = \mathrm{coeq}(\alpha,\gamma)$. The fact that $f^{\blacktriangle}\cdot\alpha = f^{\blacktriangle}\cdot\gamma$ implies that there exists $t$ satisfying $t\cdot r_f = f^{\blacktriangle}$.

$$\begin{array}{ccccc}
E_f & \overset{\alpha}{\underset{\gamma}{\rightrightarrows}} & IX^{@} & \xrightarrow{\ r_f\ } & Q_f \\
 & & & \searrow^{f^{\blacktriangle}} & \Big\downarrow t \\
 & & & & YX_{@}
\end{array}$$

Now by definition of the free run map $\mu_o^{@}$ we have $\mu_o^{@}\cdot IX^{@}\eta = \mathrm{id}_{IX^{@}}$. Again, by (21) of Section 1, we have the $\alpha$ and $\gamma$ squares of

$$\begin{array}{ccc}
E_f & \overset{\alpha}{\underset{\gamma}{\rightrightarrows}} & IX^{@} \\
{\scriptstyle E_f\eta}\Big\downarrow & & \Big\downarrow{\scriptstyle IX^{@}\eta} \\
E_fX^{@} & \overset{\alpha X^{@}}{\underset{\gamma X^{@}}{\rightrightarrows}} & IX^{@}X^{@}
\end{array}$$

each commute. Thus we have that

$$\mu_o^{@}\cdot\alpha X^{@}\cdot E_f\eta = \mu_o^{@}\cdot IX^{@}\eta\cdot\alpha = \alpha .$$

Since $f^{\blacktriangle}$, $\mu_o^{@}$ and $\alpha X^{@}$ are all dynamorphisms, it then follows that

$$f^{\blacktriangle}\cdot\mu_o^{@}\cdot\alpha X^{@} = (f^{\blacktriangle}\cdot\alpha)^{\#} .$$

But $f^{\blacktriangle}\cdot\alpha = f^{\blacktriangle}\cdot\gamma$ and so

$$f^{\blacktriangle}\cdot\mu_o^{@}\cdot\alpha X^{@} = f^{\blacktriangle}\cdot\mu_o^{@}\cdot\gamma X^{@} .$$

Then, since $(\alpha,\gamma)$ is the kernel pair of $f^{\blacktriangle}$, there exists $u$ completing

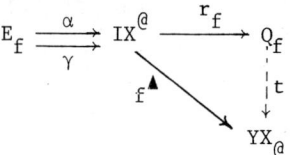

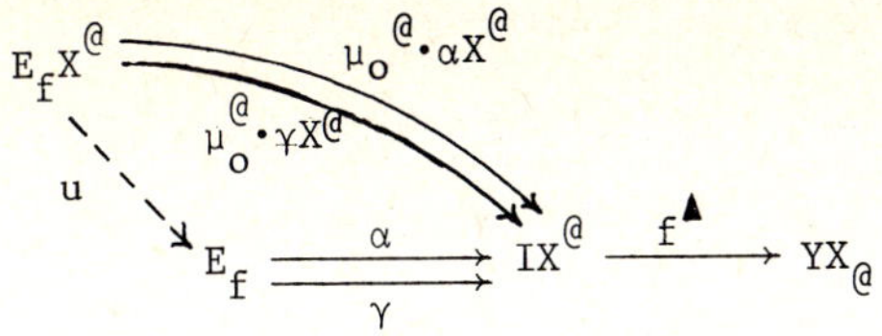

Thus $r_f \cdot \mu_o^@ \cdot \alpha X^@ = r_f \cdot \alpha \cdot u = r_f \cdot \gamma \cdot u$ since $r_f = \text{coeq}(\alpha, \gamma)$
$$= r_f \cdot \mu_o^@ \cdot \gamma X^@ .$$

Since, by our claim on $X^@$, $r_f X^@ = \text{coeq}(\alpha X^@, \gamma X^@)$ there then exists $v$ such that $v \cdot r_f X^@ = r_f \cdot \mu_o^@$

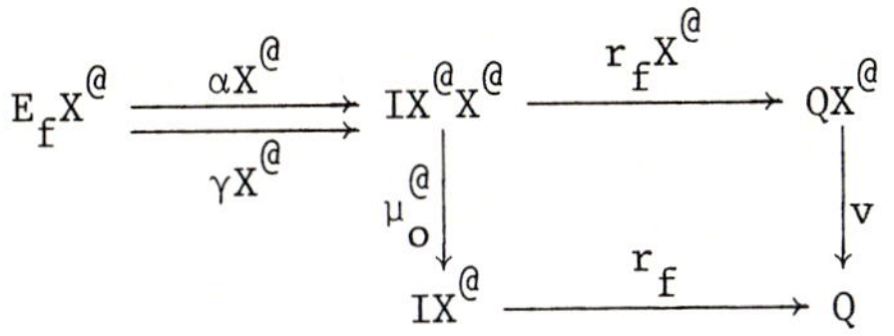

$$E_f X^@ \xrightarrow[\gamma X^@]{\alpha X^@} IX^@ X^@ \xrightarrow{r_f X^@} QX^@ \qquad (9)$$
$$\mu_o^@ \downarrow \qquad\qquad \downarrow v$$
$$IX^@ \xrightarrow{r_f} Q$$

Now by Lemma <u>1.34</u> we have the commutativity of

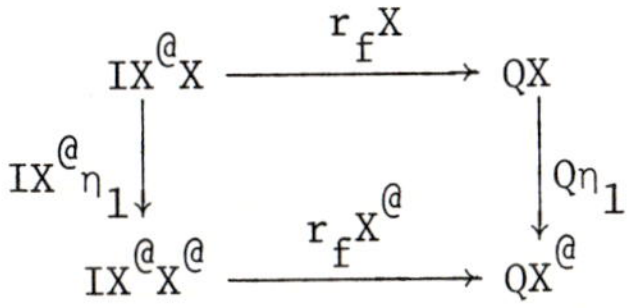

$$IX^@ X \xrightarrow{r_f X} QX$$
$$IX^@ \eta_1 \downarrow \qquad \downarrow Q\eta_1 \qquad (10)$$
$$IX^@ X^@ \xrightarrow{r_f X^@} QX^@$$

Noting that, by <u>1.36</u>, we have $\mu_o^@ \cdot IX^@ \eta_1 = \mu_o$ we may splice (10) atop (9) to obtain

$$IX^@ X \xrightarrow{r_f X} QX$$
$$\mu_o \downarrow \qquad \downarrow v \cdot Q\eta_1$$
$$IX^@ \xrightarrow{r_f} Q$$

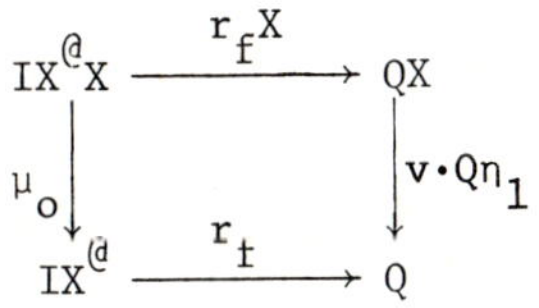

which asserts that $r_f$ is a dynamorphism $(IX^@, \mu_o) \to (Q, Q\eta_1 \cdot v)$. Thus postulate 3 is satisfied, with $\delta_f = v \cdot Q\eta_1 : QX \to Q$ the requisite dynamics of our minimal realization. □

To tie this back to the image factorizations of [2], let's see how we might attempt to construct a coequalizer-mono factorization† of $f\colon A \to B$ in $\mathcal{K}$ (suggested by the "first isomorphism theorem" of set theory): Let $p\colon A \to G = \mathrm{coeq}(t_1,t_2)$ where $(t_1,t_2)$ is the kernel pair of f. Since $ft_1 = ft_2$, there exists a unique $i\colon Q \to B$ with $f = p\cdot i$; call this the canonical factorization of f. While it is not always possible to prove that i is mono, we do have the following result which says this is the only way to construct the coequalizer-mono factorization if there is one.

2.12 FACT: Assume that the coequalizer of the kernel pair of $f\colon A \to B$ exists. Then if f has a coequalizer-mono factorization it must coincide with the canonical factorization.

Proof: Let $f = p\cdot i$ be a coequalizer-mono factorization of f, and let $(t_1,t_2)$ be the kernel pair of f. By hypothesis, we can write $p = \mathrm{coeq}(u_1,u_2)$ for some pair $(u_1,u_2)$.

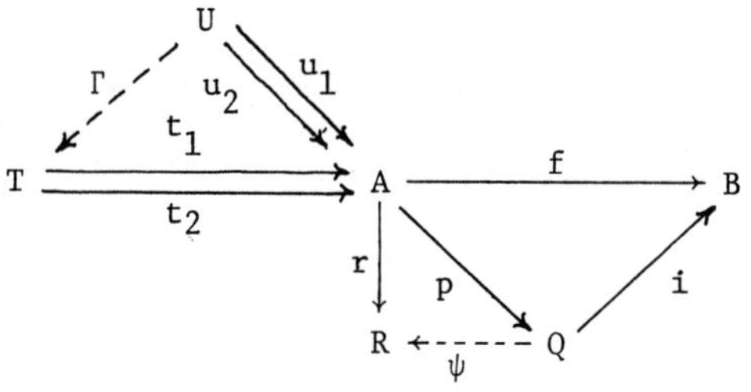


It suffices to show that $p = \mathrm{coeq}(t_1,t_2)$. Suppose $r\cdot t_1 = r\cdot t_2$. As $f\cdot u_1 = f\cdot u_2$ there exists unique $\Gamma$ with $\Gamma\cdot t_i = u_i$ since $(t_1,t_2)$ is the kernel pair of f; and it follows that $r\cdot u_1 = r\cdot u_2$ inducing a unique $\psi\colon Q \to R$ with $\psi\cdot p = r$ as desired to yield our conclusion that $p = \mathrm{coeq}(t_1,t_2)$. □

---

†It is straightforward to prove that, if every morphism in $\mathcal{K}$ may be factored as $f = i\cdot p$ with p a coequalizer and i a monomorphism, then $\mathcal{E}$ = {coequalizers}, $\mathcal{M}$ = {monomorphisms} defines an image factorization system for $\mathcal{K}$.

At the risk of minor redundancy, let us explain how 2.12 integrates with the preceding material. For state-behavior machines, we have two distinct approaches to realization:

i) Given existence of kernel pairs and coequalizers, we set up the kernel pair of $f^{\blacktriangle}$ (which we can show is the abstract Nerode equivalence 2.10), obtain its coequalizer, and note that it is a dynamorphism 2.11. A minimal realization results.

ii) Given existence of coequalizer-mono factorizations, we factor $f^{\blacktriangle}$ and note [3] that the coequalizer and mono morphisms are both dynamorphisms. A minimal realization results.

What Fact 2.12 says is that if there exist kernel pairs and coequalizer-mono factorizations, both approaches to realization yield the same thing.

Actually, even more is true, and we can connect realization theories using reflexive and abstract Nerode equivalences.

2.13 PROPOSITION: Let $\mathcal{K}$ be a category possessing coequalizers and kernel pairs. Then every coequalizer is a reflexive coequalizer.

Proof: Let $r: A \rightarrow B = \mathrm{coeq}(p_1, p_2)$ for $p_i: C \rightarrow A$. Let $t_1, t_2: D \rightarrow A$ be the kernel pair of $r$. As shown in 2.12, $r = \mathrm{coeq}(t_1, t_2)$.

It remains to show that kernel pairs are reflexive.

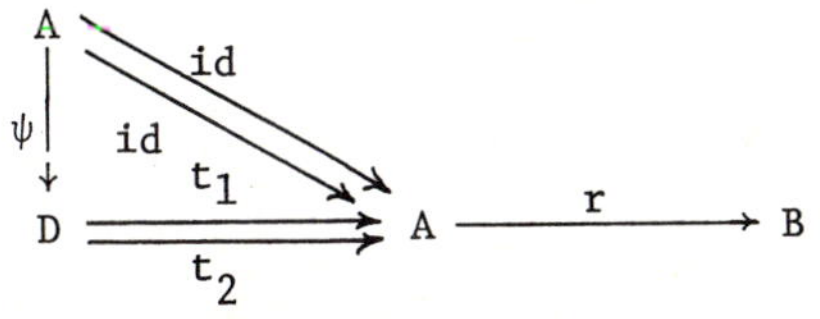


By the kernel pair property, there exists $\psi: A \rightarrow D$ making the above diagram commute. □

Now refer back to the statement of Theorem 2.6. Evidently, the abstract or reflexive Nerode equivalences together with Postulates 2 through 4, with a kernel pair existence assumption in the case of the reflexive equivalence, yield a minimal coequalizer-reachable realization of f by essentially identical construction procedures.

## 3. Tree Automata: Finite Successes and Infinite Failures.

In this section we apply the realization theory of Section 2 to a class of systems in **Set** which have proved very important in computer science: the tree automata. Since the study of tree automata plays no role in the study of reachability and observability conditions in Section 4 (and only a partial role in Section 5), many control theorists may wish to omit this section unless they have an interest in theory of computation. (An overview of applications is given in Chapter 4 of [7]).

Consider the arithmetic tree:

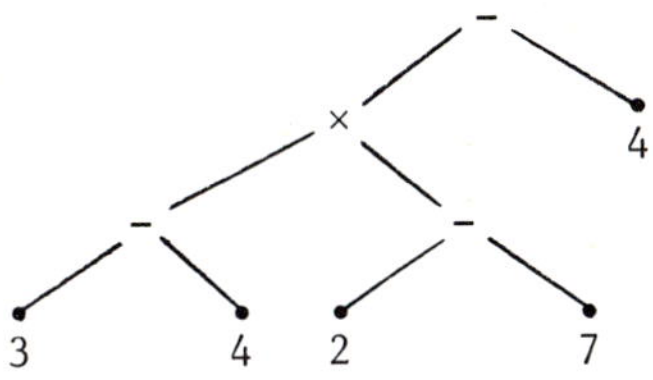


which we may regard as a representation of the arithmetic expression

$$((3 - 4) \times (2 - 7)) - 4.$$

To evaluate it we need two functions (where $\mathbb{Z}$ is the set of all integers)

$$\delta_- : \mathbb{Z} \times \mathbb{Z} \to \mathbb{Z} : (m,n) \mapsto m - n$$

$$\delta_\times : \mathbb{Z} \times \mathbb{Z} \to \mathbb{Z} : (m,n) \mapsto m \times n$$

which we may combine into a single map

$$\delta : (\mathbb{Z} \times \mathbb{Z}) + (\mathbb{Z} \times \mathbb{Z}) \to \mathbb{Z} : (m,n,\omega) \to \begin{cases} m - n & \text{if } \omega = - \\ m \times n & \text{if } \omega = \times \end{cases}$$

when we use the coproduct (disjoint union) diagrams with $in_-(m,n) = (m,n,-)$, $in_\times(m,n) = (m,n,\times)$

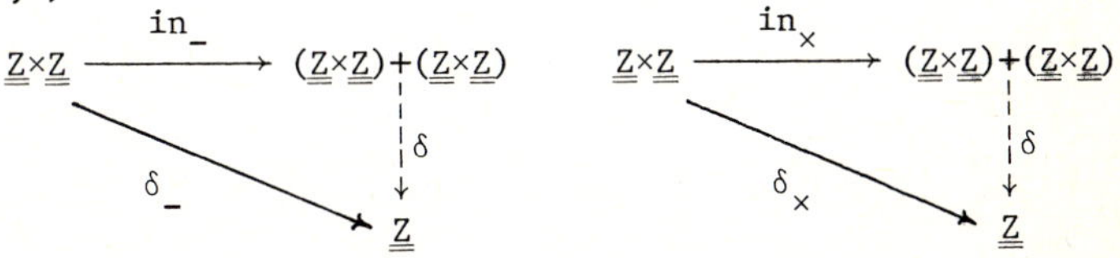

[Note that the meaning of the symbols + and × in $(\underline{\underline{Z}} \times \underline{\underline{Z}}) + (\underline{\underline{Z}} \times \underline{\underline{Z}})$ is *not* addition and multiplication, but disjoint union and cartesian product.]
We call $\{-,\times\}$ the *label set* and we say that $-$ and $\times$ are both *2-ary* labels since each labels an operator which acts on 2 arguments. With this background, the reader can appreciate that the automaton which processes whole arithmetic trees such as that shown above is a special case of the tree automata which we arrive at in the next few paragraphs:

*3.1 DEFINITION*: A *label set* is a set $\Omega$ together with a map $\nu$ which assigns to each $\omega$ in $\Omega$ a *cardinal* $\nu(\omega)$. [For much of this section, the reader may think of $\nu(\omega)$ as simply being an integer, $\nu(\omega) \in \underline{\underline{N}}$. When $\nu(\omega)$ is in $\underline{\underline{N}}$ for all $\omega$ in $\Omega$, we call $\Omega$ *finitary*.] We call $\nu(\omega)$ the *arity* of $\omega$, and set $\Omega_n$ to be the set of $\omega$ in $\Omega$ with arity n. Note that $\nu(\omega)$ may equal zero, i.e. $\Omega_o$ may be nonempty.

We shall now show how to associate an input process with each label set $\Omega$.

*3.2 DEFINITION*: Given a label set $\Omega$, we define a functor $X_\Omega : \underline{\underline{Set}} \to \underline{\underline{Set}}$ by the object mapping

$$QX_\Omega = \coprod_n \Big( \coprod_{\omega \in \Omega_n} Q^n \Big) \qquad (1)$$

while the action of $X_\Omega$ on morphisms is given by

$$\begin{array}{ccccc} Q & \quad & QX_\Omega & \xleftarrow{\ in_{n,\omega}\ } & Q^n \\ \Big\downarrow f & & \Big\downarrow fX_\Omega & \swarrow_{in_{n,\omega} f^n} & \Big\downarrow f^n \\ Q' & & Q'X_\Omega & \xleftarrow[\ in_{n,\omega}\ ]{} & (Q')^n \end{array} \qquad (2)$$

where $f^n(q_1,\ldots,q_n) = (f(q_1),\ldots,f(q_n))$.

When $n = 0$, we view $Q^n$ as a one-element set $\{1\}$, and so the only candidate for $f^0\colon Q^0 \to (Q')^0$ sends 1 to 1.

In our motivating example, $\Omega_2 = \{-,\times\}$; $\Omega_n = \emptyset$, the empty set, for $n \neq 2$. Thus

$$QX_\Omega = \coprod_{\{-,\times\}} Q^2 = Q^2\times\{-\} + Q^2\times\{\times\}$$

and $fX_\Omega : \coprod_{\{-,\times\}} Q^2 \to \coprod_{\{-,\times\}} (Q')^2$ sends $(q_1,q_2,\omega)$ to $(f(q_1),f(q_2),\omega)$.

Now just as our arithmetic tree processing required maps $\delta_-$ and $\delta_\times$, so do we obtain the more general concept:

**3.3 DEFINITION:** An $\Omega$-*algebra* is simply an $X_\Omega$-dynamics

$$\delta\colon QX_\Omega \to Q$$

which is the same thing as a collection of maps $\delta_\omega\colon Q^{\nu(\omega)} \to Q$, one for each $\omega$ in $\Omega$:

$$\begin{array}{ccc} Q^{\nu(\omega)} & \xrightarrow{\ in_{\nu(\omega),\omega}\ } & QX_\Omega \\ & \searrow^{\delta_\omega} & \downarrow{\scriptstyle \delta} \\ & & Q \end{array} \qquad (3)$$

Note that when $\nu(\omega) = 0$, $\delta_\omega$ has domain a one-element set, and so can be thought of as a particular single element of $Q$.

An $\Omega$-algebra *homomorphism* $h\colon (Q,\delta) \to (Q',\delta')$ is then an $X_\Omega$-dynamorphism, i.e. a map $h\colon Q \to Q'$ for which

$$\begin{array}{ccc} QX_\Omega & \xrightarrow{\ \delta\ } & Q \\ {\scriptstyle hX_\Omega}\downarrow & & \downarrow{\scriptstyle h} \\ Q'X_\Omega & \xrightarrow{\ \delta'\ } & Q' \end{array} \qquad (4)$$

Equivalently, for all $\omega$ in $\Omega$, with $n = \nu(\omega)$,

$$\delta'\cdot hX_\Omega\cdot in_{n,\omega} = h\cdot\delta\cdot in_{n,\omega}.$$

By 3.2, this is equivalent to

$$\delta' \cdot in_{n,\omega} \cdot h^n = h \cdot \delta \cdot in_{n,\omega}$$

and by the remark above to

$$\delta'_\omega \cdot h^n = h \cdot \delta_\omega$$

i.e.

$$\delta'_\omega(h(q_1),\ldots,h(q_n)) = h(\delta_\omega(q_1,\ldots,q_n)) \qquad (5)$$

for all $(q_1,\ldots,q_n)$ in $Q^n$, and $\omega \in \Omega_n$, whatever the choice of the cardinal n. In case $n = 0$, let $\delta_\omega$ pick out $q_0 \in Q$ and $\delta'_\omega$ pick out $q_0' \in Q'$; then (5) says that $q_0' = h(q_0)$.

We say an $\Omega$-algebra is finitary just in case $\Omega$ is finitary ($\Omega_n \neq \emptyset$ only if n finite) whether or not Q is a finite set. The theory of finitary $\Omega$-algebras is studied in [15] and [16].

**3.4 DEFINITION:** A tree automaton or $\Omega$-algebra automaton is an $X_\Omega$-system, i.e. a 7-tuple $(X_\Omega, Q, \delta, I, \tau, Y, \beta)$ where $X_\Omega$, Q and $\delta$ have the interpretations given above, $\tau: I \rightarrow Q$ is an arbitrary morphism of Set, and $\beta: Q \rightarrow Y$ is an arbitrary morphism of Set. Justification for the adjective "tree" has yet to be presented: note that I, Q and Y are not trees.

We shall now work towards establishing that $X_\Omega$ is an input process.

**3.5 DEFINITION:** Given any set I, define the set $\mathfrak{J}_{I,\Omega}$ inductively by

Basis Step: For each $i \in I$, or $\omega \in \Omega_0$, put (i) or $\omega$ in $\mathfrak{J}_{I,\Omega}$

Induction Step: For each $\omega \in \Omega_n$ and each n-tuple $(t_1,\ldots,t_n)$ already in $\mathfrak{J}_{I,\Omega}$, put $(t_1,\ldots,t_n)\omega$ in $\mathfrak{J}_{I,\Omega}$.

At this point, $(t_1,\ldots,t_n)\omega$ is a formal symbol, not the evaluation of a function; in 3.6 below, however, we reinterpret $(\cdot)\omega$ as an n-ary function on $\mathfrak{J}_{I,\Omega}$.

We call elements of $\mathcal{J}_{I,\Omega}$ $(\Omega,I)$-<u>trees</u>, or $\Omega$-<u>trees over</u> I <u>generators</u>. The reason for this terminology is best presented by a build-up of the tree that introduced this section in the following steps:

Set $\Omega_2 = \{-,\times\}$, all other $\Omega_n$ being empty. Take $I = \underline{\underline{Z}}$. The basis step provides us with the one-node trees:

$$(2) = \dot{2}\ ;\quad (3) = \dot{3}\ ;\quad (4) = \dot{4}\ ;\quad \text{and}\quad (7) = \dot{7}\ .$$

(Of course, there are other one-node trees than these.) A single application of the induction step yields two typical two-node trees:

$((3),(4))- =$ 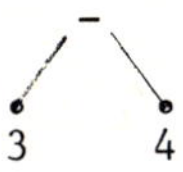

; and $((2),(7))- =$ 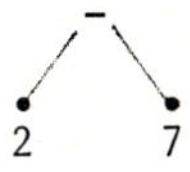


and it requires two further levels of induction to yield our tree as $((((3),(4))-,((2),(7))-)\times,(4))-$.

As an example in which $\Omega_o \neq \emptyset$, suppose $\Omega_o = \{\sqrt{-1}\}$, $\Omega_2 = \{+,\times\}$ and otherwise $\Omega_n = \emptyset$. Take $I = \underline{\underline{Z}}$, $Q = \underline{\underline{Z}} + \sqrt{-1}\,\underline{\underline{Z}}$ .

Typical one-node trees are

$$(2) = \dot{2}\ ;\quad (3) = \dot{3}\ ;\quad (4) = \dot{4}\ ;\quad \sqrt{-1} = \dot{\sqrt{-1}}$$

A single application of the induction step yields for example

$((3),(4))+ =$ 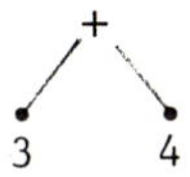

; $((2),\sqrt{-1}\ )+ =$ 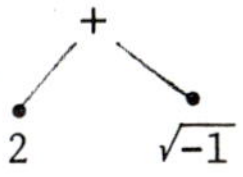


More generally, we can argue inductively; each $i \in I$ and $\omega \in \Omega_o$ determines a one-node tree:

$$\dot{(i)} \qquad\qquad \dot{\omega}$$

If $t_1,\ldots,t_n$ are all trees and $\omega \in \Omega_n$, then $(t_1,\ldots,t_n)\omega$ can be thought of as the tree

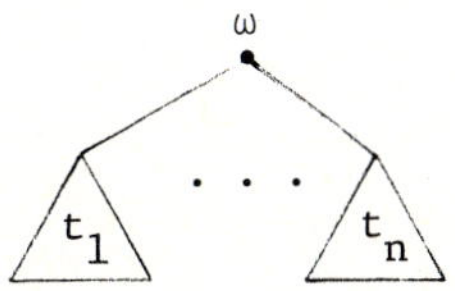

where  is shorthand for a picture of a whole tree.

Now as the picture at the foot of p. 47 suggests, <u>we may use the trees themselves to form an $\Omega$-algebra:</u>

$$\delta^{\mathfrak{J}} : \mathfrak{J}_{I,\Omega} X_\Omega \longrightarrow \mathfrak{J}_{I,\Omega}$$

with the simple composition rule that for $\omega \in \Omega_n$,

$$\delta^{\mathfrak{J}}_\omega : \mathfrak{J}_{I,\Omega}{}^n \longrightarrow \mathfrak{J}_{I,\Omega} : (t_1,\ldots,t_n) \mapsto (t_1,\ldots,t_n)\omega$$

which is certainly a valid map by the induction step in the definition of $\mathfrak{J}_{I,\Omega}$. Let us check that this yields the free $X_\Omega$-dynamics in the sense that $\mathfrak{J}_{I,\Omega}$ corresponds to $I(X_\Omega)^@$ and $\delta^{\mathfrak{J}}$ corresponds to the morphism $\mu_o : I(X_\Omega)^@ X_\Omega \to I(X_\Omega)^@$. With slight abuse of notation, write $IX_\Omega^@$ for $I(X_\Omega)^@$.

<u>3.6</u> <u>THEOREM</u>: Let $\Omega$ be a label set. Then for any set I, the $X_\Omega$-dynamics

$$\mu_o : (IX_\Omega^@)X_\Omega \longrightarrow IX_\Omega^@$$

with $IX_\Omega^@ = \mathfrak{J}_{I,\Omega}$ and $\mu_o = \delta^{\mathfrak{J}}$ is free over I. In other words, given any $\Omega$-algebra $(Q,\delta)$, and any map $\tau : I \to Q$, there exists a unique homomorphism $\psi : (IX_\Omega^@, \mu_o) \to (Q,\delta)$ such that

$$\begin{array}{ccc} I & \xrightarrow{I\eta} & IX_\Omega^@ \\ & \searrow^{\tau} & \downarrow{\scriptstyle\psi} \\ & & Q \end{array} \qquad \begin{array}{ccc} (IX_\Omega^@)X_\Omega & \xrightarrow{I\mu_o} & IX_\Omega^@ \\ {\scriptstyle\psi X_\Omega}\downarrow & & \downarrow{\scriptstyle\psi} \\ QX_\Omega & \xrightarrow{\delta} & Q \end{array} \qquad (6)$$

where the "inclusion of generators" $I\eta : I \to IX_\Omega^@$ views $i \in I$ as the "one-node tree" $\eta(i) = (i) \in \mathfrak{J}_{I,\Omega} = IX_\Omega^@$.

Proof: We simply use the inductive definition of $\mathfrak{J}_{I,\Omega}$ to specify $\psi$. The triangle says

$$\psi(t) = \tau(i) \quad \text{if} \quad t = (i) \quad \text{for } i \in I, \qquad (7)$$

while the square says that if $\psi(t_1),\ldots,\psi(t_n)$ are already defined, then for any $\omega$ in $\Omega_n$ we must have, by the equivalence of (4) and (5),

$$\psi((t_1,\ldots,t_n)\omega) = \delta_\omega(\psi(t_1),\ldots,\psi(t_n)). \tag{8}$$

In this way, $\psi$ is both well-defined and uniquely defined, with (6) holding. □

Now we can see the reason for the name tree automaton: any element of $IX_\Omega^@$ is a tree, and $\psi$ describes how this tree is processed to yield an element of Q. More precisely, we use $\tau$ to relabel the initial nodes (i) with $\tau(i) \in Q$ and $\omega$ **with** $\delta_\omega$ when $\nu(\omega) = 0$ **to relabel the initial nodes with elements** of Q--for example

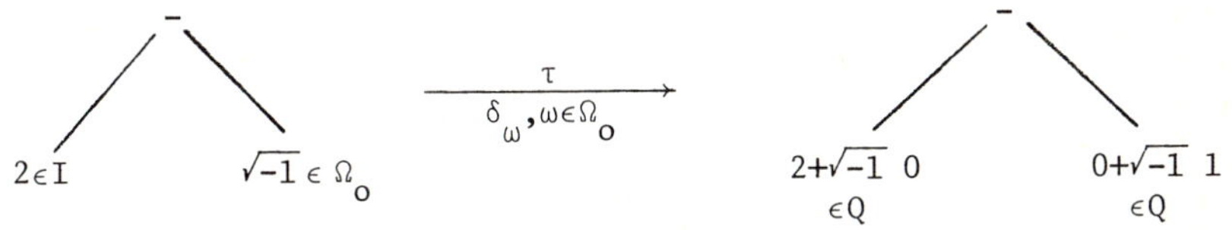


Then one "runs" $(Q,\delta)$ on the tree, passing from "leaves" to "root" and applying the appropriate $\delta_\omega$. For example

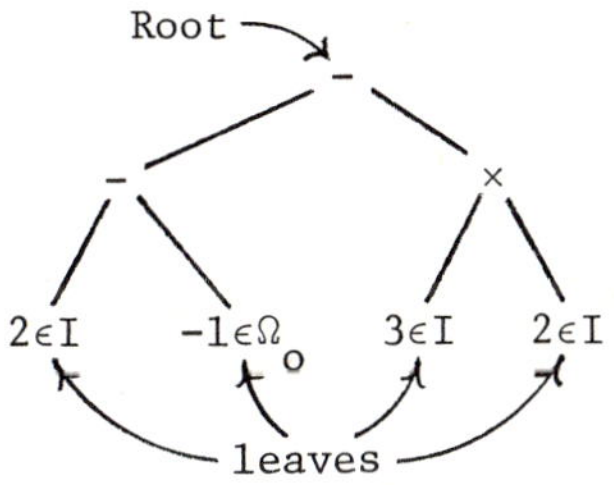


yields

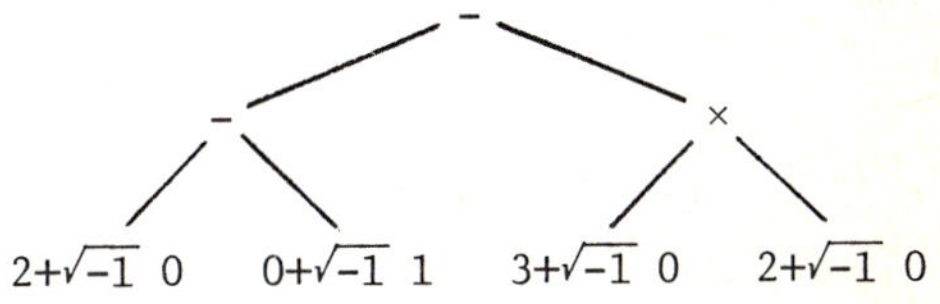


and then

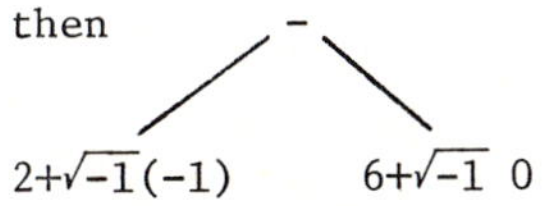


and then $-4 + \sqrt{-1}\ (-1)$.

The mapping $\psi$ associates with any element of $\mathfrak{J}_{I,\Omega}$ an element $q \in Q$.

It is interesting to note that $\psi$ can be built up in another way. Let $\tau X^@_\Omega : IX^@_\Omega \to QX^@_\Omega$ be the mapping taking a tree in $IX^@_\Omega$ and relabelling the leaves by changing $i \in I$ to $\tau(i) \in Q$ (note that it is only at [not necessarily all of] the leaves that elements of I appear--at the other nodes, operator labels appear). In this way then, $\tau X^@_\Omega$ yields a tree in $QX^@_\Omega$. Also, let $\delta^*: QX^@_\Omega \to Q$ be the unique dynamorphic extension of $id_Q$ (obtained by taking $\tau = id_Q : Q \to Q$ in (6)). Then we have, by an easy induction argument, that

$$\psi = \delta^* \cdot \tau X^@_\Omega : IX^@_\Omega \to QX^@_\Omega \to Q$$

which generalizes the familiar formula $\psi(q,w) = \delta^*(\tau(q),w)$ of sequential machine theory.

<u>3.7</u> <u>DEFINITION:</u> A subset R of $Q \times Q$ is a <u>congruence</u> on the $\Omega$-algebra $(Q,\delta)$ if it is an equivalence relation which is also a <u>subalgebra of</u> $(Q\times Q, \delta\times\delta)$; i.e.

$$\omega\in\Omega_n, \text{ and } (q_i,q_i')\in R \text{ for } i\in n \Longrightarrow (\delta_\omega(q_1,\ldots,q_n),\delta_\omega(q_1',\ldots,q_n'))\in R. \tag{9}$$

Note that (9) is automatic for $n = 0$ since it asserts $(\delta_\omega,\delta_\omega)\in R$ if $\omega\in\Omega_0$.

<u>3.8</u> <u>OBSERVATION:</u> Let $Q/R$ be the set of equivalence classes of Q with respect to a congruence R. Then the assignment

$$\bar{\delta} : (Q/R)X_\Omega \to (Q/R)$$

defined by

$$\bar{\delta}_\omega : (Q/R)^{\nu(\omega)} \to (Q/R) : ([q_1],\ldots,[q_n]) \mapsto [\delta_\omega(q_1,\ldots,q_n)]$$

provides the unique $\Omega$-algebra structure on $Q/R$ such that

$$\pi: Q \to Q/R : q \mapsto [q]$$

is an $\Omega$-homomorphism:

$$\begin{array}{ccc} QX_\Omega & \xrightarrow{\delta} & Q \\ \pi X_\Omega \downarrow & & \downarrow \pi \\ (Q/R)X_\Omega & \xrightarrow{\bar{\delta}} & Q/R \end{array}$$

The proof is an immediate consequence of the definition. $\square$

<u>3.9</u> <u>OBSERVATION</u>: Let $\psi\colon (Q,\delta) \to (Q',\delta')$ be a dynamorphism. Then $\psi(Q) = \{\psi(q) \mid q \in Q\}$ is a subalgebra of $(Q',\delta')$, i.e. given $\omega \in \Omega_n$ and $(q'_1,\ldots,q'_n) \in (\psi(Q))^n$, $\delta'_\omega(q'_1,\ldots,q'_n) \in \psi(Q)$. Also, $E = \{(q_1,q'_1) \mid \psi(q_1) = \psi(q'_1)\}$ is a congruence of $(Q,\delta)$.

Proof: Recalling (4) and (5), we have $\delta'_\omega(q'_1,\ldots,q'_n) = \delta'_\omega(\psi(q_1),\ldots,\psi(q_n))$ $= \psi(\delta_\omega(q_1),\ldots,\delta_\omega(q_n))$, proving the first claim.

To prove the second claim, observe that E is clearly an equivalence relation. Suppose that $(q_i,q'_i) \in E$ for $i = 1,2,\ldots,n$. For arbitrary $\omega \in \Omega_n$, $\psi(\delta_\omega(q_1),\ldots,\delta_\omega(q_n)) = \delta_\omega{}'(\psi(q_1),\ldots,\psi(q_n)) = \delta_\omega{}'(\psi(q'_1),\ldots,\psi(q'_n))$ $= \psi(\delta_\omega(q'_1),\ldots,\delta_\omega(q'_n))$ and this proves the subalgebra, and hence congruency, property. $\square$

<u>3.10</u> <u>DEFINITION</u>: Let $(Q,\delta) \in \mathrm{Dyn}(X_\Omega)$. If $j \in \bar{n} = \{1,2,\ldots,n\}$ and if $q_i \in Q$ are given for all $i \in \bar{n}$ with $i \neq j$ define $(q_i;j,q) \in Q^n$ by

$$(q_i;j,q) = (r_1,r_2,\ldots,r_n), \qquad r_i = \begin{cases} q_i & i \neq j \\ q & i = j\,. \end{cases}$$

Thus, if $n = 4$ and $j = 3$, $(q_i;j,q) = (q_1,q_2,q,q_4)$. An <u>elementary translation</u> of $(Q,\delta)$ is a function $\tau\colon Q \to Q$ for which there exist $n > 0$, $\omega \in \Omega_n$, $j \in \bar{n}$ and fixed $q_i \in Q$ for all $i \in n$, $i \neq j$ such that

$$\tau(q) = \delta_\omega(q_i;j,q)\ . \qquad (10)$$

[For example, if a vector space Q is regarded as an $\Omega$-dynamics with (among other things) + in $\Omega_2$, $q \mapsto q + q_0$ is an elementary translation for each fixed $q_0$.]

A <u>translation of</u> $(Q,\delta)$ is an element of the submonoid of $Q^Q$ generated by the elementary translations; i.e., $\tau: Q \to Q$ is a translation if $\tau = id_Q$ or if $\tau$ is a composition $\tau_n \circ \dots \circ \tau_1$ of elementary translations.

For each set I the $\Omega$-<u>monoid</u>, $M_\Omega(I)$, of I is the monoid of translations of $(IX_\Omega^@, \mu_o)$.

<u>3.11 EXAMPLE</u>: Let $\Omega_1 = X_o$ (i.e. each $x \in X_o$ is the label of a unary operator), $\Omega_n = \emptyset$ for $n \neq 1$. Then $QX_\Omega = Q \times X_o$, i.e. sequential machine theory is a particular case of tree automata theory. An element $(i, x_1 \dots x_n)$ of $IX^@ = I \times X_o^*$ is the tree

$$\begin{array}{c} x_n \\ | \\ \vdots \\ | \\ x_1 \\ | \\ i \end{array}$$

Every elementary translation $\tau: IX^@ \to IX^@$ has the form $\delta_x$ for some $x \in X_o$ with $\delta_x(i,w) = (i,wx)$. The general translation is then $(i,w) \mapsto (i,ww')$. In case $I = 1$, $M(1) = \{R_{w'} \mid w' \in X_o^*\}$, where $R_w: X_o^* \to X_o^*$ is the "right translation" $w \mapsto ww'$.

We recalled, at the top of p. 28, that the classical Nerode equivalence for $- \times X_o$: <u>Set</u> $\to$ <u>Set</u> has

$$(w_1,w_2) \in E_f \iff f(w_1 w) = f(w_2 w) \quad \text{for all} \quad w \in X_o^*$$

which may now be rewritten as

$$(w_1,w_2) \in E_f \iff fR_w w_1 = fR_w w_2 \quad \text{for all} \quad w \in X_o^*.$$

Thus the classical Nerode equivalence of the response $f: X_o^* \to Y$ is seen to be

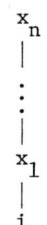

$$E_f = \{(w_1,w_2) \mid f\tau w_1 = f\tau w_2 \quad \text{for all } \tau \in M(1)\}.$$

But this construction works for finitary $X_\Omega$:

3.12 THEOREM: Let $X = X_\Omega$ for a finitary label set $\Omega$, let $f\colon IX_\Omega^@ \to Y$ be a function. Define

$$E_f = \{(x,y) \in IX_\Omega^@ \times IX_\Omega^@ \mid f\tau x = f\tau y \quad \text{for all } \tau \in M_\Omega(I)\}.$$

Then $E_f$ is a congruence of $(IX_\Omega^@, I\mu_o)$ and

$$\alpha,\gamma\colon E_f \to IX_\Omega^@, \quad \text{where} \quad \alpha(x,y) = x \quad \text{and} \quad \gamma(x,y) = y$$

is the reflexive Nerode equivalence of f.

Postulates 2, 3, 4 of theorem 2.6 are satisfied and so (recalling 2.13) the coequalizer-minimal realization $M_f$ of f exists with the canonical projection $r_f\colon IX_\Omega^@ \to (IX_\Omega^@/E_f) = Q_f$ as the reachability map.

Proof: $E_f$ is obviously an equivalence relation. We must show that if $(x_1,y_1),\ldots,(x_n,y_n) \in E_f$ and if $\omega \in \Omega_n$ then $((x_1,\ldots,x_n)\omega,(y_1,\ldots,y_n)\omega) \in E_f$. This is clear if $n = 0$ since it is true that $(\omega_o,\omega_o) \in E_f$. If $n > 0$, the result follows (and here we make crucial use of the fact that n is finite!) from the following chain of equalities in which $\tau \in M_\Omega(I)$ is arbitrary:

$$\begin{aligned} & f\tau((x_1,x_2,\ldots,x_n)\omega) \\ &= f\tau((y_1,x_2,\ldots,x_n)\omega) \\ &= f\tau((y_1,y_2,\ldots,x_n)\omega) \\ & \quad \cdots \\ &= f\tau((y_1,\ldots,y_{n-1},x_n)\omega) \\ &= f\tau((y_1,\ldots,y_{n-1},y_n)\omega). \end{aligned}$$

To see why the first equality holds, observe that

$$\tau'\colon a \mapsto (a,x_2,\ldots,x_n)\omega$$

is an elementary translation. Hence $\tau\tau'$ is a translation, with $\tau\tau'(x_1) = \tau((x_1,x_2,\ldots,x_n)\omega)$ and $\tau\tau'(y_1) = \tau((y_1,x_2,\ldots,x_n)\omega)$. Because $(x_1,y_1) \in E_f$, $f\tau\tau'(x_1) = f\tau\tau'(y_1)$, i.e. the first equality holds. The other equalities are obtained similarly.

Now that $E_f$ is a congruence we may use 3.8 to construct $(Q_f,\delta_f) = IX^@_\Omega/E_f$ with canonical projection $r_f$

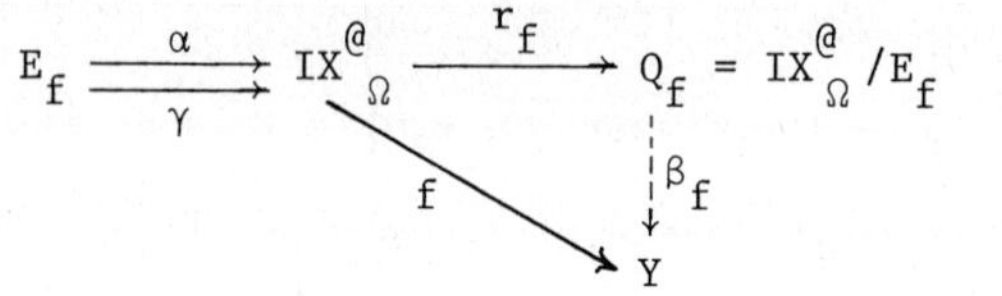

(11)

First, we shall prove $f\alpha^\# = f\gamma^\#$. Since $r_f = \mathrm{coeq}(\alpha,\gamma)$ in Set and $f\alpha = f\gamma$ (i.e. consider $\tau = \mathrm{id}$ in the definition of $E_f$), there exists unique $\beta_f$ with $\beta_f r_f = f$. Therefore, $f\alpha^\# = \beta_f r_f \alpha\# = \beta_f(r_f\cdot\alpha)^\#$ [i.e. $r_f\alpha^\#$, $(r_f\alpha)^\#$ are both dynamorphisms equal to $r_f\alpha$ when preceded by $E_f\eta$ and hence are equal by 3.6] $= \beta_f(r_f\cdot\gamma)^\# = f\gamma^\#$. Now, to show that $E_f$ is the reflexive Nerode equivalence, suppose $t,u\colon T \to IX^@_\Omega$ is another reflexive pair satisfying $ft^\# = fu^\#$. To induce the desired $\Gamma$ (d will be explained below)

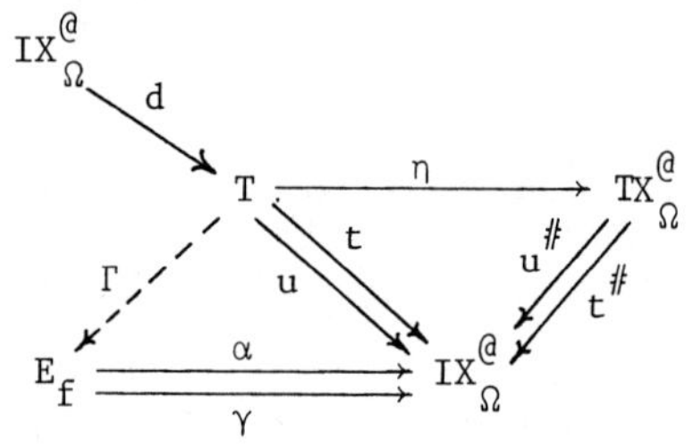

as shown above it is necessary and sufficient to show that $\{(t(a),u(a)) \mid a \in T\}$ is a subset of $E_f$. (Of course, $E_f \subset IX^@_\Omega \times IX^@_\Omega$.) Since $t = t^\#\eta$, $u = u^\#\eta$, one has $\mathrm{Im}\,t \subset \mathrm{Im}\,t^\#$ and $\mathrm{Im}\,u \subset \mathrm{Im}\,u^\#$; it then suffices to show that $T_1 = \{(t^\#(b),u^\#(b) \mid b \in TX^@_\Omega\}$ is a subset of $E_f$, i.e. that $f\tau t^\#(b) = f\tau u^\#(b)$ for all $b \in TX^@_\Omega$, $\tau \in M(I)$.

To this end we pause to observe that $T_1$ is reflexive, i.e. $(x,x) \in T_1$ for all $x \in IX^@_\Omega$; for, there exists d with $td = id = ud$ and hence $t^\#(\eta d) = id = u^\#(\eta d)$. Also, $\mathrm{Im}\, t^\#$ and $\mathrm{Im}\, u^\#$ by 3.9 are subalgebras of $IX^@_\Omega$, so that $T_1 = \mathrm{Im}\, t^\# \times \mathrm{Im}\, u^\#$ is a subalgebra of $IX^@ \times IX^@$. Hence, if $b \in TX^@_\Omega$ and $\tau$ is the elementary translation $x \mapsto (x_1,\ldots,x_{j-1},x,x_{j+1},\ldots,x_n)\omega$ then, because $(x_1,x_1),\ldots,(x_{j-1},x_{j-1}),(t^\#(b),u^\#(b)),(x_{j+1},x_{j+1}),\ldots,(x_n,x_n) \in T_1$, $(\tau(t^\#(b)),\tau(u^\#(b))) \in T_1$ by the subalgebra property. By iteration, $(\tau(t^\#(b)),\tau(u^\#(b))) \in T_1$ for any $\tau \in M(I)$.

Now let $(x,y) \in T_1$; then $(\tau x,\tau y) \in T_1$ by what we have proved for all $\tau \in M(I)$. Thus $\tau x = t^\#(b)$, $\tau y = u^\#(b)$ for some $b \in TX^@_\Omega$. But by assumption, $ft^\# = fu^\#$, so that $f\tau x = f\tau y$ and then $(x,y) \in E_f$, i.e. $T_1 \subset E_f$. This completes the proof that $\alpha,\gamma\colon E_f \longrightarrow IX^@_\Omega$ is the reflexive Nerode equivalence of f.

The rest is easy. Postulates 2 and 3 of 2.6 have already been established as $r_f$ and $\delta_f$ above and postulate 4 holds for any $X\colon \underline{\underline{\mathrm{Set}}} \longrightarrow \underline{\underline{\mathrm{Set}}}$ for the following reason. Observe that for any epi $\alpha\colon A \longrightarrow B$ in $\underline{\underline{\mathrm{Set}}}$, there is a readily constructed $\beta\colon B \longrightarrow A$ with $\alpha\cdot\beta = id_B$. Then $\alpha X\cdot\beta X = id_{BX}$, so that $\alpha X$ is an epi. □

In previous literature on realization theory for tree automata [17] it has been the practice to consider $f\colon IX^@{}_\Omega \longrightarrow Y$ only for $I = \emptyset$. A justification for this practice was that if $I \neq \emptyset$ we may define a new label set $\Omega(I)$ by

$$(\Omega(I))_0 = \Omega_0 + I,$$

$$(\Omega(I))_n = \Omega_n \quad \text{for} \quad n > 0$$

and then observe that there is a canonical bijection

$$\emptyset X^@{}_{\Omega(I)} \cong IX^@{}_\Omega .$$

Since each element of $\emptyset X^{@}_{\Omega}$ is a "0-ary derived operation" [15], it follows that elementary translations--hence all translations--on $\emptyset X^{@}$ are derived unary operations. This observation is the basis of the Nerode equivalence formula

$$E_f = \{(x,y) \in \emptyset X^{@}_{\Omega} \times \emptyset X^{@}_{\Omega} \mid f\tau x = f\tau y \text{ for all derived unary operations } \tau \text{ on } IX^{@}_{\Omega}\}$$

found (in different notation) as [17, 6.9]. Theorem 3.12 above shows that in general the relevant $\tau$'s are translations, not derived operations. The $\Omega(I)$ approach is not always natural. For example, in a study of machine interconnection--where the output set of one machine is the input set of another--it is natural to fix $X_{\Omega}$ and let I vary.

3.13 EXAMPLE: $E_f$ as in 3.12 need not be the abstract Nerode equivalence: let $X = X_{\Omega}$ where $\Omega_2 = \{\omega\}$, $\Omega_n = \emptyset$ for $n \neq 2$. Let $I = \{a,b\}$, $Y = \{0,1\}$ and define $f\colon IX^{@} \longrightarrow Y$ by

$$f(x) = \begin{cases} 1 \text{ if } x = (ab)\omega \\ 0 \text{ if } x \neq (ab)\omega \end{cases}$$

Define $E_f$ as in 3.12 and define $t,u\colon T \longrightarrow IX^{@}_{\Omega}$ with

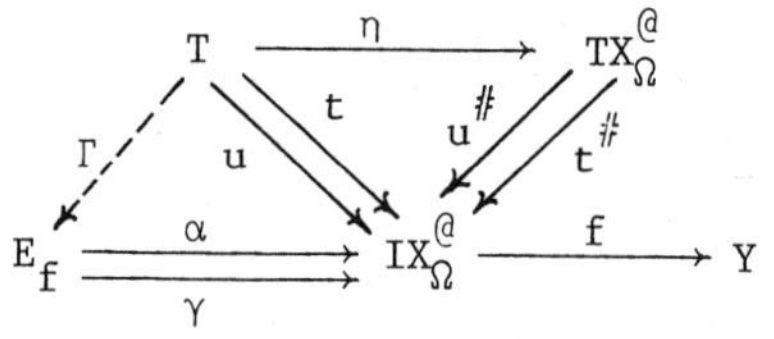


by $T = \{(a,(aa)\omega)\}$, $t(a,(aa)\omega) = a$, $u(a,(aa)\omega) = (aa)\omega$. It is clear by induction (as in 3.5) that the images of $t^{\#}$, $u^{\#}$ in $IX^{@}$ contain no trees in which b appears as a leaf and, in particular, $ft^{\#} = fu^{\#}$.

To show that $\Gamma$ does not exist it is necessary and sufficient to find $\tau \in M(I)$ such that $f\tau(a) \neq f\tau((aa)\omega)$. Take $\tau$ defined by $x \mapsto (xb)\omega$. Then

$f\tau(a) = f((ab)\omega) = 1$, while $f\tau((aa)\omega)) = f(((aa)\omega b)\omega) = 0$. Of course, $t,u\colon T \to IX^{@}_{\Omega}$ is not reflexive.

We now give compelling evidence as to why the reflexive Nerode equivalence is of wider applicability than the external Nerode equivalence of 2.2:

3.14 THEOREM: Let $\Omega$ be finitary and contain at least one operation of arity $\geq 2$. Let Y have 2 elements. Then for any $I \neq \emptyset$ there exists $f\colon IX^{@}_{\Omega} \to Y$ such that postulate 3 of 2.6--the existence of a minimal dynamics--with respect to the external Nerode equivalence does not hold (although it does hold for the reflexive Nerode equivalence of 2.1 by 3.12).

Proof: Recall that $\mu^{(n)}\colon IX^{@}_{\Omega}X^{n}_{\Omega} \to IX^{@}_{\Omega}$ is defined by

$$\mu^{(0)} = id$$
$$\mu^{(n+1)} = IX^{@}_{\Omega}X^{n}_{\Omega}X_{\Omega} \xrightarrow{\mu^{(n)}X_{\Omega}} IX^{@}_{\Omega}X_{\Omega} \xrightarrow{\delta^{J}} IX^{@}_{\Omega} .$$

Set $D_n = Im(\mu^{(n)}) \subset IX^{@}_{\Omega}$. Suppose, for illustration, that there exists $\omega \in \Omega_2$, $a,b,c \in I$. Then $ab\omega \in D_1$, $ab\omega ac\omega\omega \in D_2$, but $ab\omega c\omega \notin D_n$ for any n. $D_1$ coincides with trees of the form , $D_2$ with trees of the form  (i.e. trees obtained by replacing the leaves of a $D_1$ tree with a whole $D_1$ tree); $D_3$ coincides with trees obtained by replacing the leaves of a $D_1$ tree by whole $D_2$ trees, and so on. Notice that the tree for $ab\omega c\omega$ is

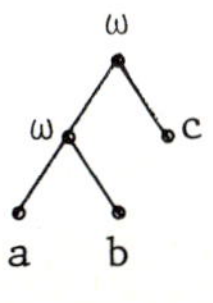


which will not look like a $D_n$ tree for any n.

Set $D = \bigcup_{n=1}^{\infty} D_n$ and define

$$f\colon IX^{@}_{\Omega} \to Y = \chi_D.$$

Set $E \underset{\gamma}{\overset{\alpha}{\rightrightarrows}} IX^{@}_{\Omega}$ to be the kernel pair of f.

We claim that E is the external Nerode equivalence of f. For clearly it satisfies $f\alpha = f\gamma$, i.e. $f\cdot\mu^{(o)}\cdot\alpha X^{o} = f\cdot\mu^{(o)}\cdot\gamma X^{o}$. For all $n \geq 1$, $\mathrm{Im}(\mu^{(n)}\cdot\alpha X^{n})$, $\mathrm{Im}(\mu^{(n)}\cdot\gamma X^{n}) \subset \mathrm{Im}(\mu^{(n)}) = D_n \subset D$, so that $f\cdot\mu^{(n)}\cdot\alpha X = f\cdot\mu^{(n)}\cdot\gamma X^{n}$ for all $n \geq 0$. Further if $E' \underset{\gamma'}{\overset{\alpha'}{\rightrightarrows}} IX^{@}_{\Omega}$ is such that $f\cdot\mu^{(n)}\cdot\alpha' X^{n} = f\cdot\mu^{(n)}\cdot\gamma' X^{n}$ for all $n \geq 0$, taking $n = 0$ and using the kernel pair property yields a unique $\psi$ completing

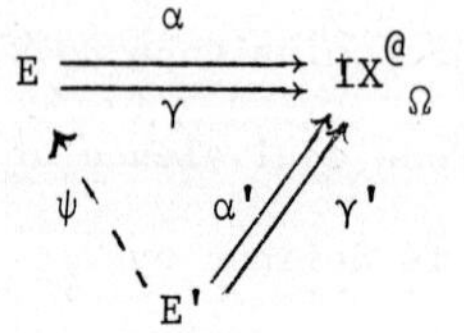


Now let $\omega$ be any operation of arity $> 1$ and set $n = \nu(\omega)$. Let $a \in I$. Define

$$P_1 = (a,\ldots,a)\omega \in D_1$$

$$P_2 = (P_1,\ldots,P_1)\omega \in D_2$$

Then $(P_1,P_1) \in E$ and $(P_1,P_2) \in E$ but $((P_1,\ldots,P_1,P_1)\omega,\ (P_1,\ldots,P_1,P_2)\omega) \notin E$ as $P_2 = (P_1,\ldots,P_1,P_1)\omega \in D_2 \subset D$, $(P_1,\ldots,P_1,P_2)\omega \notin D$.

Thus $E \underset{\gamma}{\overset{\alpha}{\rightrightarrows}} IX^{@}_{\Omega}$ is not a congruence. It follows from 3.9 that $IX^{@}_{\Omega} \xrightarrow{r} Q = \mathrm{coeq}(\alpha,\gamma)$ carries no $\Omega$-structure making r a homomorphism, so postulate 3 fails as asserted. □

Consider the label set $\Omega_2 = \{\omega\}$, $\Omega_n = \emptyset$ $n \neq 2$. Consider the inclusions $j_n \colon IX^{n}_{\Omega} \rightarrow IX^{@}_{\Omega}$. The image of $j_n$ is the set of <u>homogeneous formulas of degree</u> n. [Thus $(((a,b)\omega,(c,d)\omega)\omega)$ is homogeneous of degree 2.] The downfall of the external Nerode equivalence is that not all formulas are homogeneous. This difficulty can be surmounted by extending $\Omega$ to $\overline{\Omega}$ by adding a unary operation

$\Delta$, called "unit boost". $\Omega$-formulas are now representable as homogeneous $\bar{\Omega}$ formulas. To illustrate,

$$((a,b)\omega,(c,d)\omega)\omega \longmapsto ((a,b)\omega,(c,d)\omega)\omega$$

$$((a,b)\omega,c)\omega \longmapsto ((a,b)\omega,(c)\Delta)\omega$$

$$[\{(a,b)\omega,((c,d)\omega,e)\omega\}\omega,(x,y)\omega]\omega \longrightarrow [\{(<a>\Delta,<b>\Delta)\omega,((c,d)\omega,<e>\Delta)\omega\}\omega,(<<x>\Delta>\Delta,<<y>\Delta>\Delta)\omega]\omega$$

Clearly $QX_{\bar{\Omega}} = QX_{\Omega} + Q$. This suggests the idea of augmenting the input process, which we shall take up in Section 5.

We conclude this section with some remarks about infinitary label sets. There are many examples of infinitary $\Omega$-algebras such as lattices with varying degrees of completeness, Boolean $\sigma$-rings (as used in measure theory), commutative $C^*$-algebras and so on. See [18, chapter 1, section 5]. Even though the minimal realization problem for infinitary tree automata has no immediate application to problems of either control theory or computer science it is worth our while to show that, in the infinitary case, not every f has a minimal realization; for this demonstrates that "existence of minimal realizations" is not a consequence of completeness and cocompleteness (see [10, chap. III]) of Dyn(X) and $\mathcal{K}$ , and strongly suggests that "existence of minimal realizations" is a "finiteness condition" on X.

3.15 LEMMA: Let $\Omega$ be a label set, $X = X_\Omega$ and let I be a set. Then to construct a response $f: IX^@ \to Y$ with no minimal realization it suffices to construct a sequence $(R_n)$ of congruences on $IX^@$ such that $R_n \subset R_{n+1}$ for all n and such that $R = \cup R_n$ is not a subalgebra of $IX^@ \times IX^@$.

Proof: Since $R_n \subset R_{n+1}$, R is an equivalence relation. Set $Y = IX^@/R$ and let $f: IX^@ \to Y$ be the canonical projection. Suppose that f has a minimal realization $M_f$ as shown below but that R is not a subalgebra. We shall deduce a contradiction.

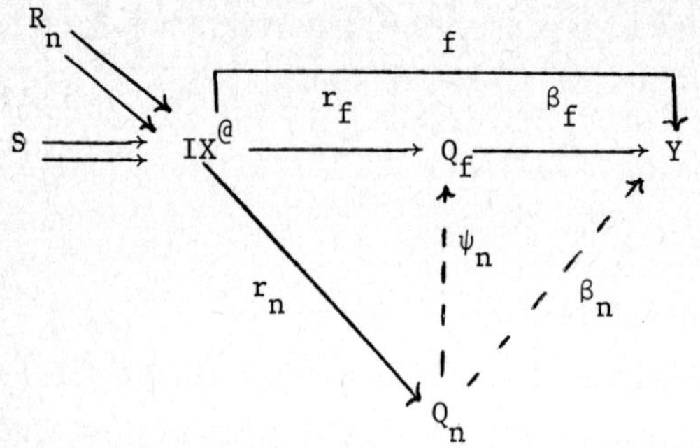


Define $S = \{(x,y) \mid r_f(x) = r_f(y)\}$. Then S is a congruence by 3.9. Set $Q_n = IX^@/R_n$ with canonical projection $r_n$ and let $\delta_n : Q_nX \to Q_n$ be the unique dynamics (see 3.8) such that $r_n$ is a dynamorphism. Since $R_n \subset R$ there exists unique $\beta_n$ with $\beta_n r_n = f$. Since $M_f$ is minimal there must exist a unique $\psi_n$ with $\psi_n r_n = r_f$ and $\beta_f \psi_n = \beta_n$. Let $(x,y) \in R$. Then $(x,y) \in R_n$ for some n. Hence $r_f(x) = \psi_n r_n(x) = \psi_n r_n(y) = r_f(y)$, i.e. $(x,y) \in S$. Conversely, let $(x,y) \in S$. Then $f(x) = \beta_f r_f(x) = \beta_f r_f(y) = f(y)$, so $(x,y) \in R$. Since $R = S$ and S is a subalgebra of $IX^@ \times IX^@$, this contradicts the assumption about R. □

The following theorem builds on this Lemma to exhibit an $f: IX^@ \to Y$ with no minimal realization.

3.16 THEOREM: Let $\alpha$ be the countably infinite cardinal, let $\Omega$ be the infinitary label set with $\Omega_\alpha = \{\omega\}$, $\Omega_n = \emptyset$ if $n \neq \alpha$, let $X = X_\Omega$ and let $I = \{0,1,2,\ldots\}$. Then there exists Y and $f: IX^@ \to Y$ such that f has no minimal realization.

Proof: Define $h: I \to I$ by

$$h(i) = \begin{cases} i - 1 & \text{if} \quad i \geq 1 \\ 0 & \text{if} \quad i = 0 \end{cases}$$

Then $h^n(i) = 0$ if $0 \leq i \leq n$ and $h^n(n + k) = k$. Let $\psi: IX^@ \to IX^@ = hX^@$. Specifically, recall that $IX^@$ is inductively defined by

(basis step) $i \in IX^@$ (if $i \in I$)

(inductive step) If $P_k \in IX^@$ for all $k \in \alpha$, $(P_k)\omega \in IX^@$.

(Here $(P_k)$ is short hand for the sequence $P_1, P_2, \ldots)$. Thus $\psi$ is inductively defined by

$$\psi(i) = h(i)$$

$$\psi((P_k)\omega) = (\psi(P_k))\omega$$

Note that $\psi^n = (hX^@)^n = h^n X^@$ is inductively defined by

$$\psi^n(i) = h^n(i)$$

$$\psi^n((P_k)\omega) = (\psi^n(P_k))\omega$$

Define $R_n = \{(x,y) \mid \psi^n(x) = \psi^n(y)\}$. Then $R_n$ is a congruence on $IX^@$ by 3.9. If $\psi^n(x) = \psi^n(y)$ then $\psi(\psi^n(x)) = \psi(\psi^n(y))$, so $R_n \subset R_{n+1}$. By 3.15, it suffices to show $R = \cup R_n$ is not a subalgebra; to do this we will define $P_n \in IX^@$ so that $(P_0, P_n) \in R$ but $((P_0P_0P_0\ldots)\omega, (P_0P_1P_2\ldots)\omega) \notin R$. Accordingly, let

$$P_n = (012\ldots n000\ldots)\omega$$

Then $\psi^n(P_n) = (h^n(0)h^n(1)\ldots h^n(n)h^n(0)h^n(0)\ldots)\omega$

$$= (000\ldots)\omega = P_0 = \psi^n(P_0)$$

So $(P_0, P_n) \in R_n \subset R$. If $((P_0P_0P_0\ldots)\omega, (P_0P_1P_2\ldots)\omega) \in R$ we must have

$$\psi^n((P_0P_0P_0\ldots)\omega) = \psi^n((P_0P_1P_2\ldots)\omega) \tag{12}$$

for some n. But

$$\psi^n((P_0P_0P_0\ldots)\omega) = (\psi^n(P_0)\psi^n(P_0)\ldots)\omega$$

$$= (P_0P_0\ldots)\omega$$

whereas

$$\psi^n((P_0P_1P_2\ldots)\omega) = (\psi^n(P_0)\psi^n(P_1)\psi^n(P_2)\ldots)\omega$$

If (12) holds then we must have

$$P_0 = \psi^n(P_k)$$

for every k. But

$$\psi^n(P_{n+1}) = (h^n(0)h^n(1)\ldots h^n(n)h^n(n+1)h^n(0)h^n(0)\ldots)\omega$$
$$= (0\ldots010\ldots0)\omega$$
$$\neq (00\ldots)\omega = P_0$$

This exhibits the desired contradiction. □

The question of when $f : IX^@ \to Y$ has a minimal realization has been considered by Adámek [22] and Trnková [23, 24]. The most powerful 'non-existence theorem' at this writing is [24, proposition 3]:

> Suppose there exists an infinite cardinal $\alpha \leq \text{card}(IX^@)$ for which there exists p in $\alpha X$ such that p is not in the image of fX for any $f : A \to \alpha$ with $\text{card}(A) < \alpha$. Then there exists a subset of $IX^@$ whose characteristic function has no minimal realization.

The hypotheses of this theorem hold for the X of 3.16 with $\alpha = \aleph_0$ thereby showing that the statement of 3.16 can be considerably strengthened. Trnková's construction, even for this X, is rather complicated, and we have failed to construct an example with $I = 1$, $Y = 2$ that compares in simplicity to 3.16.

## 4. Finite Step Conditions.

Arbib and Zeiger [9] in giving a unified (but "pre-categorical") view of linear systems and sequential machines, observed that the subspaces

$$S_1 \subset S_2 \subset S_3 \subset \ldots$$

with $S_j$ being the set of states reachable in at most k steps from the initial state had the property that if ever $S_k$ equalled $S_{k+1}$, then all the $S_j$'s were equal for $j \geq k$. They observed that whenever $S_j$ was not equal to $S_{j+1}$, then $S_{j+1}$ exceeded $S_j$ by at least 1 in cardinality for sequential machines, and by at least 1 in dimensionality for linear systems. They could then exhibit the commonality of the results that for a sequential machine of n states the set of reachable states was $S_{n-1}$, while for a linear system of dimension n the space of reachable states was $S_n$. We now provide a general categorical setting for these results, and the dual observability results. In this section, we provide a theory applicable, in particular, to adjoint processes (which include our classic examples of sequential machines and linear systems); while Section 5 gives a more general theory applicable to tree automata, which are not even state-behavior.

Until further notice in this section we assume that our category $\mathcal{K}$ has finite coproducts, and an image factorization system (see Section 3 of [2], [11] or [12]). By the theory of and following 1.33 we have:

4.1 FACT: Given an input process X, and a system $M = (X,Q,\delta,I,\tau,Y,\beta)$, the initial state $\tau: I \to Q$ has dynamorphic extension the reachability map $r: IX^@ \to Q$, and (by Section 1 (20) with $a = \tau$) this satisfies

$$r \cdot \eta_k = \delta^{(k)} \cdot \tau X^k : IX^k \to Q. \qquad \square$$

Given the linear system $G: I \to Q$, $F: Q \to Q$, $H: Q \to Y$ ($\mathcal{K}$ = Vect, $X = \mathrm{id}_{\mathcal{K}}$), $r \cdot \eta_k$ sends u to $F^k Gu$; an input string (...0 u 0...0) where u occurs at time -k takes the zero state to the state $F^k Gu$ at time 0.

For $X = - \times X_o$ : Set $\to$ Set, $r \cdot \eta_k : I \times X_o^k \to Q$ sends $(q_o, w)$ to $\delta^*(q_o, w)$ the state reachable from initial state $q_o$ by applying the input string w of length k. Since the coproduct $(in_k : IX^k \to \coprod_{j=0}^{n-1} IX^j)$ gives us the object of "less-than-n-step" input applications, we see the usefulness of the definition:

4.2 DEFINITION: The less-than-n-steps-reachability map of a system with reachability r is the map $r_n : \coprod_{j=0}^{n-1} IX^j \to Q$ given by $in_k \cdot r_n = \delta^{(k)} \cdot \tau X^k$.

Define $in_o^{n-1} : \coprod_{j=0}^{n-1} IX^j \longrightarrow \coprod_{j=0}^{n} IX^j$ by

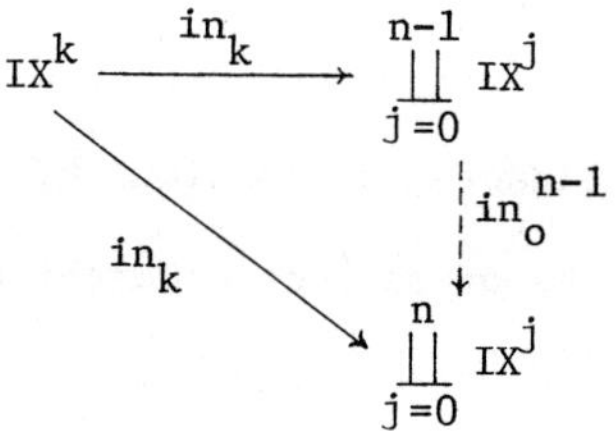


$in_o^{n-1} \cdot in_k = in_k$

(The diagram should make clear the two different uses of the symbol $in_k$.)

Then

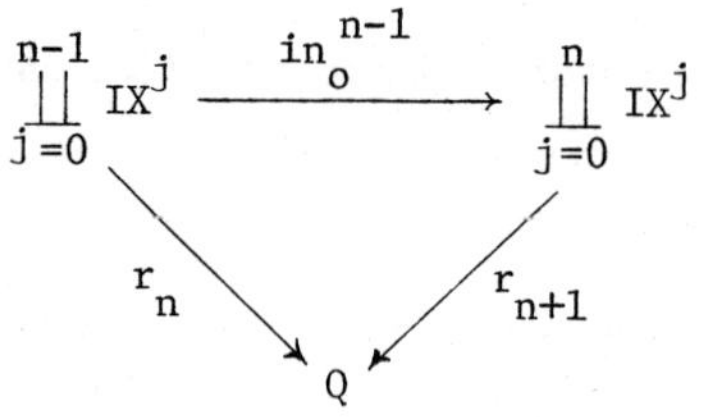


(1)

commutes, since for $0 \le k < n$, $r_{n+1} \cdot in_o^{n-1} \cdot in_k = r_{n+1} \cdot in_k = \delta^{(k)} \cdot \tau X^k = r_n \cdot in_k$.

Thus we may apply diagonal fill in (See Lemma 3.5 of [2]; or [11], [12]) to obtain:

<u>4.3</u> <u>THEOREM</u>: Let each $r_n$ have $\mathcal{E}$-$\mathcal{M}$ factorization $r_n = m_n \cdot e_n$. Then there exists a unique $h_{n+1}$ in $\mathcal{M}$ such that

$$\begin{array}{ccccc} \coprod_{j=0}^{n-1} IX^j & & \xrightarrow{\quad in_0^{n-1} \quad} & & \coprod_{j=0}^{n} IX^j \\ & {\scriptstyle e_n}\searrow & & \swarrow{\scriptstyle e_{n+1}} & \\ & Q^{\textcircled{n}} & \overset{h_{n+1}}{- - - \rightarrow} & Q^{\textcircled{n+1}} & \\ & {\scriptstyle m_n}\searrow & & \swarrow{\scriptstyle m_{n+1}} & \\ & & Q & & \end{array} \qquad (2)$$

commutes. □

In our classic examples, this says that $Q^{\textcircled{n}}$ is, up to isomorphism, a subset of $Q^{\textcircled{n+1}}$, and so captures the existence of the chain $\ldots S_n \subset S_{n+1} \subset \ldots$ noted above. The next result then captures the idea that if we can ever reach the whole state-space ($r_n$ is onto in our classic examples; $r_n \in \mathcal{E}$ in our general theory) then we can always reach it all thereafter:

<u>4.4</u> <u>COROLLARY</u>: If $r_n \in \mathcal{E}$, then $r_{n+1} \in \mathcal{E}$, and $Q^{\textcircled{k}} \cong Q$ for all $k \geq n$.

Proof: Lemma 5.6 of [2] (or [11]) tells us that whenever $gf \in \mathcal{E}$, we must have $g \in \mathcal{E}$. Then $r_{n+1} \in \mathcal{E}$ follows from (1), and so $r_k \in \mathcal{E}$ for all $k \geq n$. But if $r_k$ is in $\mathcal{E}$, then $m_k$ in its $\mathcal{E}$-$\mathcal{M}$ factorization must be an isomorphism, and $Q^{\textcircled{k}} \cong Q$. In fact, we can choose $r_k = e_k$, $m_k = id_Q$, to force $Q^{\textcircled{k}} = Q$. □

In the rest of this section we develop our theory for processes X which preserve finite coproduct diagrams (hence, by <u>1.31</u> and <u>1.26</u>, we include adjoint processes, and, with them, linear systems and sequential machines).

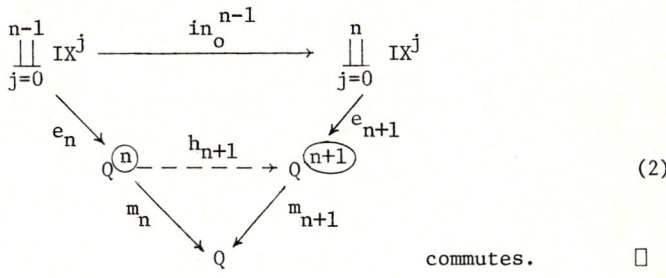

We may thus apply X to the coproduct diagram $in_j \colon IX^j \to \coprod_{j=0}^{n-1} IX^j$ in order to define

$$\mu_{n,1} \colon \Big(\coprod_{j=0}^{n-1} IX^j\Big) X \longrightarrow \coprod_{j=0}^{n} IX^j$$

from the coproduct diagram

$$\begin{array}{ccc}
\Big(\coprod_{j=0}^{n-1} IX^j\Big)X & \overset{\mu_{n,1}}{\dashrightarrow} & \coprod_{j=0}^{n} IX^j \\
{\scriptstyle in_j X}\big\downarrow & \nearrow{\scriptstyle in_{j+1}} & \\
IX^j X & &
\end{array}$$

For example, for $X = - \times X_0 \colon \underline{\underline{Set}} \to \underline{\underline{Set}}$, $\mu_{n,1}$ is simply the map $(w,x) \mapsto wx$ sending a string w of length n, and a letter $x \in X_0$, to the string wx of length $n+1$. [Incidentally, by <u>1.28</u>, if X preserves countable coproducts, and $\mathcal{K}$ has countable coproducts, then $IX^@ = \coprod_{k \geq 0} IX^n$.]

<u>4.5</u> <u>THEOREM</u>: Let $\mathcal{K}$ have finite coproducts, and an image factorization system $(\mathcal{E},\mathcal{M})$. Let the input process X preserve finite coproducts and $\mathcal{E}$-morphisms. Then the following diagram defines a unique $k_{n+1} \colon Q^{(n)}X \to Q^{(n+1)}$.

$$\begin{array}{ccc}
\Big(\coprod_{j=0}^{n-1} IX^j\Big)X & \xrightarrow{\mu_{n,1}} & \coprod_{j=0}^{n} IX^j \\
{\scriptstyle e_n X}\big\downarrow & & \big\downarrow{\scriptstyle e_{n+1}} \\
Q^{(n)}X & \overset{k_{n+1}}{\dashrightarrow} & Q^{(n+1)} \\
{\scriptstyle m_n X}\big\downarrow & & \big\downarrow{\scriptstyle m_{n+1}} \\
QX & \xrightarrow{\delta} & Q
\end{array} \qquad (3)$$

(NOTE: The passage from $\delta$ to $(k_{n+1})$ describes a functor from Dyn(X) to the category of time-varying X-dynamics is the sense of [19]).

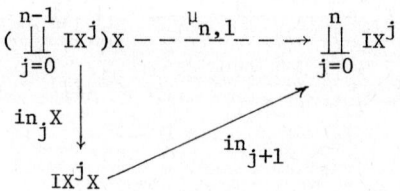

Proof: To check commutativity of the outer part of the diagram, we use the fact that $\coprod_{j=0}^{n-1} (IX^j)X$ is a coproduct to reduce our checking to showing that

$$r_{n+1}\cdot\mu_{n,1}\cdot in_j X \quad = \quad \delta\cdot r_n X\cdot in_j X, \qquad \text{for } 0 \le j < n.$$

But $r_{n+1}\cdot(\mu_{n,1}\cdot in_j X) = r_{n+1}\cdot in_{j+1}$ by the definition of $\mu_{n,1}$

$= \delta^{(j+1)}\cdot\tau X^{j+1}$ by definition 4.2

$= \delta\cdot\delta^{(j)}X\cdot\tau X^{j+1}$ by the definition of $\delta^{(j+1)}$

$= \delta\cdot(\delta^{(j)}\cdot\tau X^j)X$ since X is a functor

$= \delta\cdot(r_n\cdot in_j)X$ by 4.2, again

$= \delta\cdot r_n X\cdot in_j$

That $k_{n+1}$ exists follows by the diagonal fill-in lemma, since $e_n X \in \mathcal{E}$ and $m_{n+1} \in \mathcal{M}$. □

In our classical examples, $k_{n+1}$ is simply the map that acts on a state reachable in less than n steps to provide a state reachable in less than n+1 steps. We now come to the main result which may be paraphrased "if one step gets you no further then no finite number of steps can take you further: when you stick, you're stuck!"

4.6 THEOREM: Under the conditions of 4.5; and with the definitions of 4.3, if $h_n$ is an isomorphism then $h_{n+1}$ is an isomorphism, and hence $h_{n+k}$ is an isomorphism for all $k \ge 0$.

Proof: We shall show that if $h_n$ is an isomorphism, then we may define $v_{n+1}$ as shown in the diagram.

$$\begin{array}{ccc} \coprod_{j=0}^{n-1} IX^j & \xrightarrow{\ in_o^{n-1}\ } & \coprod_{j=0}^{n} IX^j \\ {\scriptstyle e_n}\big\downarrow & \overset{v_{n+1}}{\swarrow} & \big\downarrow{\scriptstyle e_{n+1}} \\ Q^{\textcircled{n}} & \xrightarrow[\ h_{n+1}\ ]{} & Q^{\textcircled{n+1}} \end{array} \tag{4}$$

As soon as such a $v_{n+1}$ exists, so that $h_{n+1}v_{n+1} = e_{n+1} \in \mathcal{E}$, it follows that $h_{n+1} \in \mathcal{E}$. But $h_{n+1}$ is in $\mathcal{M}$ by 4.3, and so $h_{n+1}$ is an isomorphism.

Assuming that $h_n$ is an isomorphism, so that $h_n^{-1}$ exists, we may define the morphism

$$k_n \cdot h_n^{-1} X \cdot e_n X : \quad (\coprod_{j=0}^{n-1} IX^j)X \to Q^{\textcircled{n}}$$

where we have applied X to the diagram

$$\begin{array}{ccc} \coprod_{j=0}^{n-2} IX^j & \xrightarrow{\ in_o^{n-2}\ } & \coprod_{j=0}^{n-1} IX^j \\ {\scriptstyle e_{n-1}}\big\downarrow & & \big\downarrow{\scriptstyle e_n} \\ Q^{\textcircled{n-1}} & \underset{h_n^{-1}}{\overset{h_n}{\rightleftarrows}} & Q^{\textcircled{n}} \end{array}$$

and followed $(h_n^{-1} \cdot e_n)X$ with $k_n : Q^{\textcircled{n-1}}X \to Q^{\textcircled{n}}$. Letting $in_o$ be the injection $I \to \coprod_{j=0}^{n-1} IX^j$ we also define

$$e_n \cdot in_o : I \to Q^{\textcircled{n}}.$$

Since we clearly have a coproduct diagram

$$I \xrightarrow{\ in_o\ } \coprod_{j=0}^{n} IX^j \xleftarrow{\ \mu_{n,1}\ } (\coprod_{j=0}^{n-1} IX^j)X$$

these two morphisms combine uniquely to yield as our candidate for the desired morphism,

$$v_{n+1} = \begin{pmatrix} e_n \cdot in_o \\ \hline k_n \cdot h_n^{-1} X \cdot e_n X \end{pmatrix} : \coprod_{j=0}^{n} IX^j \to Q^n.$$

It only remains to check that (4) commutes. To do this we shall only verify that

$$h_{n+1} \cdot v_{n+1} = e_{n+1}$$

since the other triangle (which is irrelevant to our theorem anyway) then commutes as $h_{n+1} \in \mathcal{M}$. As $m_{n+1} \in \mathcal{M}$, it suffices to verify that

$$m_{n+1} \cdot h_{n+1} \cdot v_{n+1} = m_{n+1} \cdot e_{n+1},$$

i.e. not only that

$$m_{n+1} \cdot h_{n+1} \cdot e_n \cdot in_o = r_{n+1} \cdot in_o,$$

[which is immediate since $m_{n+1} \cdot h_{n+1} \cdot e_n = m_n \cdot e_n$ by (2), and since $r_n \cdot in_o = r_{n+1} \cdot in_o$] but also that

$$m_{n+1} \cdot h_{n+1} \cdot k_n \cdot h_n^{-1} X \cdot e_n X = r_{n+1} \cdot \mu_{n,1}. \tag{5}$$

But

$$\begin{aligned} m_{n+1} \cdot h_{n+1} \cdot k_n \cdot h_n^{-1} X \cdot e_n X &= m_n \cdot k_n \cdot h_n^{-1} X \cdot e_n X && \text{by (2)} \\ &= \delta \cdot m_{n-1} X \cdot h_n^{-1} X \cdot e_n X && \text{by (3)} \\ &= \delta \cdot (m_{n-1} \cdot h_n^{-1} \cdot e_n) X && \\ &= \delta \cdot r_n X && \text{by (2)} \\ &= r_{n+1} \cdot \mu_{n,1} && \text{by (3)} \end{aligned}$$

which establishes (4), and with it, our theorem. □

Since $\mathcal{E}$-$\mathcal{M}$ factorization is only defined up to isomorphism, we may, in the above circumstances, set

$$\bar{Q} = Q^{(n-1)} = Q^{(n)} = \ldots = Q^{(n+k)} = \ldots$$

forcing each $h_{n+k}$, $k \geq 0$, to be the identity $id_{\bar{Q}}$. It then follows from (2) that $$m_{n+k} = m_{n-1} = \bar{m}, \quad \text{say} \qquad \text{for all } k \geq 0.$$

Thus

$$\bar{m} \cdot k_{n+\ell} = \delta \cdot \bar{m}X \qquad \text{for all } \ell \geq 0;$$

and since $\bar{m}$ is mono, it follows that the $k_{n+\ell}$'s are all equal, say to $\bar{\delta}$. Then

$$\bar{m} \cdot \bar{\delta} = \delta \cdot \bar{m}X. \tag{6}$$

By repeated application of 4.3, we obtain a monomorphism

$\hat{h} = h_n \cdot \ldots \cdot h_1 \cdot h_o : Q^{(o)} \longrightarrow \bar{Q}$ such that $\bar{m} \cdot \hat{h} = m_o$, yielding

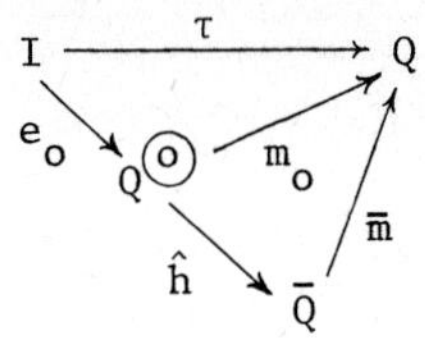


Now consider the system with dynamics $\bar{\delta} : \bar{Q}X \longrightarrow \bar{Q}$ and I-frame $\bar{\tau} = \hat{h} \cdot e_o : I \longrightarrow \bar{Q}$. Let its reachability map be $\bar{r} : IX^@ \longrightarrow \bar{Q}$. We then have the diagram

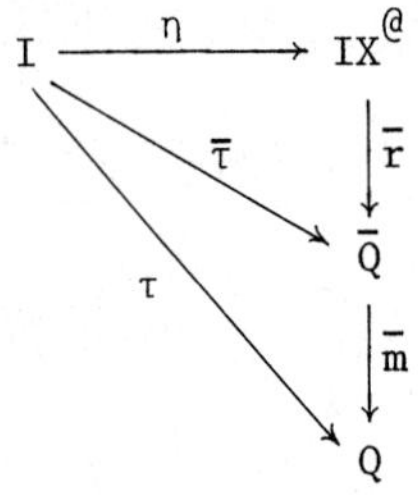


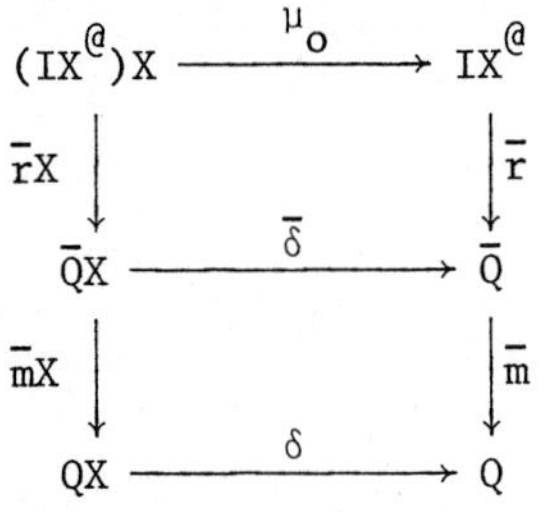


(Recall (6) in checking the commutativity.)

Thus, since $\tau$ has a unique dynamorphic extension, we have

$$r = \bar{m} \cdot \bar{r}.$$

But from (3),

$$r = \bar{m} \cdot e_{n+1},$$

and since $\bar{m} \in \mathcal{M}$, it follows that $\bar{r} = e_{n+1} \in \mathcal{E}$. Hence we have shown:

<u>4.7</u> <u>THEOREM</u>: Let $\mathcal{K}$ have finite coproducts; and let the input process X preserve finite coproducts and $\mathcal{E}$-morphisms. If $h_n$ as defined in (2) for the pair $(\delta: QX \to Q,\ \ \tau: I \to Q)$ is an isomorphism then, setting $\bar{m} = h_{n-1}$, $\bar{\delta} = k_n$, and $\bar{\tau} = h_n \cdot \ldots \cdot h_o \cdot e_o$, we have that $(\bar{\delta}: \bar{Q}X \to \bar{Q},$ $\bar{\tau}: I \to \bar{Q})$ is reachable, with reachability map $\bar{r}: IX^{@} \to \bar{Q}$ in $\mathcal{E}$ satisfying $\bar{m} \cdot \bar{r} = r$. That is, $(\bar{\delta},\bar{\tau})$ is the "reachable part" of $(\delta,\tau)$. □

In particular, we reiterate that if X is an adjoint process (so that $\mathcal{K}$ has countable coproducts and products, and X has a right adjoint) then X preserves coproducts. An adjoint process also preserves $\mathcal{E}$-morphisms for most popular choices of $\mathcal{E}$. Thus <u>4.6</u> and <u>4.7</u> apply to adjoint processes. Let us now sketch the observability theory for adjoint processes (we assume the reader familiar with the $(X^{\cdot})^{op}$ duality of [3]):

<u>4.8</u> <u>DEFINITION</u>: The dual of $r_n : \coprod_{j=0}^{n-1} IX^j \to Q$ yields the <u>at-most-n-steps observability map</u>

$$\sigma_n : Q \to \prod_{j=0}^{n-1} Y(X^{\cdot})^j .$$

[It is clear that this approximates $\sigma: Q \to YX_{@} = \prod_{j\geq 0} Y(X^{\cdot})^j$.] Then the dual of (2) for $(X^{\cdot})^{op}$ yields a unique $s_{n+1}$ in $\mathcal{E}$ such that

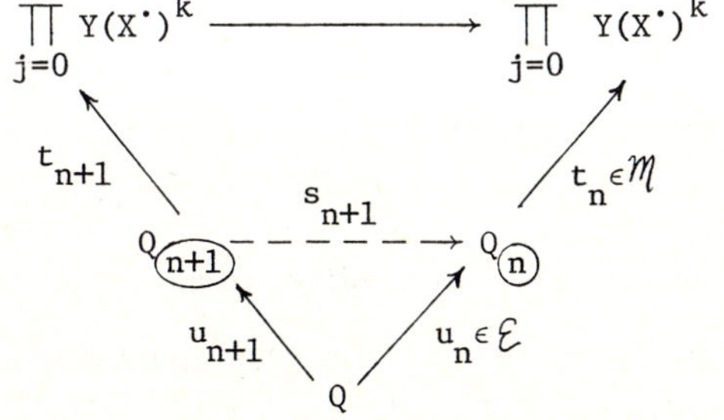


(7)

commutes. For sequential machines, $Q_n$ is the set of equivalence classes of states distinguishable at the output by an input string of length at most n; for linear systems, $Q_n$ is the set of equivalence classes of states distinguishable by studying output sequences of length n.

The dual of (3) for $(X^{\cdot})^{op}$ yields a unique $\ell_{n+1}$ such that

$$\begin{array}{ccc} [\prod_{j=0}^{n} Y(X^{\cdot})^{j}]X & \longrightarrow & \prod_{j=0}^{n-1} Y(X^{\cdot})^{j} \\ \uparrow t_{n+1} & & \uparrow t_n \\ Q_{(n+1)}X & \overset{\ell_{n+1}}{\dashrightarrow} & Q_{(n)} \\ u_{n+1} \uparrow & & \uparrow u_n \\ QX & \xrightarrow{\delta^{\cdot}} & Q \end{array} \qquad (8)$$

and in place of 4.7 we obtain

4.9 THEOREM: Let $\mathcal{K}$ have finite products; and let X be an output process that preserves products and morphisms. Then if $s_n$ as defined in (7) for the pair $(\delta: QX \to Q, \ \beta: Q \to Y)$ is an isomorphism then, setting $\bar{Q} = Q_{(n-1)}$, $\bar{u} = u_{n-1}$, $\bar{\delta} = \ell_o$ and $\bar{\beta} = t_o \cdot s_o \cdot \ldots \cdot s_{n-1}$ we have that $(\bar{\delta}: \bar{Q}X \to Q; \ \bar{\beta}: \bar{Q} \to Y)$ is an observable pair with observability map $\bar{\sigma}: \bar{Q} \to YX_@$ in $\mathcal{M}$ satisfying $\bar{\sigma} \cdot \bar{u} = \sigma$. □

Let us consider the special case of decomposable systems, so that $X = id$ and

$$r_n = \begin{pmatrix} G \\ FG \\ \vdots \\ F^{n-1}G \end{pmatrix} . \qquad (9)$$

In linear system theory, the finite-dimensionality of a state-space can reflect itself in the condition $F^nG = r_n\alpha$ for some morphism $\alpha$. To generalize this situation, we now introduce two new concepts, with the speculation that they will be useful in other contexts.

4.10 DEFINITION: Let $\mathcal{K}$ have an image factorization system $(\mathcal{E},\mathcal{M})$. We say an object A has $\mathcal{E}$-height $\ell$ over B if there exists a chain

$$A \xrightarrow{e_1} A_1 \xrightarrow{e_2} A_2 \longrightarrow \cdots \xrightarrow{e_\ell} B$$

of $\ell$ $\mathcal{E}$-morphisms none of which is an isomorphism while any such chain of length $\ell+1$ must contain an isomorphism. We say A has $\mathcal{E}$-height $\ell$ if $\ell$ is the maximum $\mathcal{E}$-height of A over any $\mathcal{K}$-object B.

Dually, we define $\mathcal{M}$-height by considering chains of $\mathcal{M}$-morphisms

$$B \xrightarrow{m_\ell} B_{\ell-1} \longrightarrow \cdots \xrightarrow{m_2} B_1 \xrightarrow{m_1} A.$$

4.11 EXAMPLES

(i) In the category Vect of vector spaces and linear maps, the $\mathcal{E}$-height of $\mathbb{R}^m$ over $\mathbb{R}^n$ is defined iff $m \geq n$, in which case it is $m-n$. The $\mathcal{E}$-height of $\mathbb{R}^m$ is m. $\mathcal{M}$-height yields the same numbers, and thus both heights correspond to dimension.

(ii) In the category Set, the $\mathcal{E}$-height of A over B is defined iff $|A| \geq |B|$, where the cardinalities $|A|$ and $|B|$ are finite and if we do not have $A \neq \phi$ but $B = \phi$. It is then $|A| - |B|$. If $\infty > |A| > 0$, then A has $\mathcal{E}$-height $|A| - 1$. $\mathcal{M}$-height yields the same numbers, and--apart from a 1--both heights correspond to cardinality.

Then 4.5 and 4.9 yield:

4.12 THEOREM: Let $\mathcal{K}$ have finite coproducts, and an image factorization system $(\mathcal{E},\mathcal{M})$. Let the input process X preserve coproducts and $\mathcal{E}$-morphisms. Let $M = (X,Q,\delta,I,\tau,Y,\beta)$ be a reachable X-system. Then

$$Q \simeq Q^{\textcircled{n_1}}$$

where $n_1$ is the $\mathcal{M}$-height of Q over I.

Dually, let $\mathcal{K}$ have finite products, and let the output process X preserve products and $\mathcal{M}$-morphisms. Let $M = (X,Q,\delta,I,\tau,Y,\beta)$ be an observable X-system. Then

$$Q \simeq Q_{\textcircled{n_2}}$$

when $n_2$ is the $\mathcal{E}$-height of Q over Y. □

**4.13 DEFINITION**: An object A of a category $\mathcal{K}$ is termed <u>projective</u> if given an arbitrary $\theta: B \to C \in \mathcal{E}$ and $\gamma: A \to C$, there exists $\alpha$ making the following diagram commute

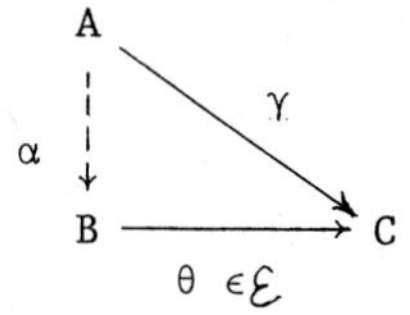


In <u>R-Mod</u>, free modules are projective, as are retracts of free objects in <u>R-Mod</u>; similarly for <u>Grp</u>. In <u>Vect</u>, all objects are projective.

We now generalize the linear system results as follows:

**4.14 THEOREM**: Let $\mathcal{K}$ be a category with products, coproducts and $\mathcal{E}$-$\mathcal{M}$ factorization, and suppose $G: I \to Q$, $F: Q \to Q$ and $H: Q \to Y$ define a decomposable system with I projective. With $r_n$ as defined in 4.2, $h_{n+1}$ as in (2), $h_{n+1}$ is an isomorphism if and only if for some $\alpha: I \to nI$, $F^nG = r_n\alpha$.

Proof: Suppose $F^nG = r_n\alpha$. Let $\tilde{e}_{n+1} : (n+1)I \to Q^{\textcircled{n}}$ be the coproduct mapping of $e_n : nI \to Q^{\textcircled{n}}$ and $e_n\alpha : I \to Q^{\textcircled{n}}$. Then

$$m_n \cdot \tilde{e}_{n+1} = \begin{pmatrix} m_n e_n \\ m_n e_n \alpha \end{pmatrix} = \begin{pmatrix} r_n \\ r_n \alpha \end{pmatrix} = \begin{pmatrix} r_n \\ F^n G \end{pmatrix} = r_{n+1} \qquad \text{by (9).}$$

Further, $\tilde{e}_{n+1} \in \mathcal{E}$ since $e_n \in \mathcal{E}$. So $m_n \cdot \tilde{e}_{n+1}$ defines an $\mathcal{E}$-$\mathcal{M}$ factorization of $r_{n+1}$, whence $Q^{(n)}$ is isomorphic to $Q^{(n+1)}$. [Notice that for this part of the proof, the projective property was not used.] Conversely, suppose $h_{n+1}$ is an isomorphism. From (2), we have

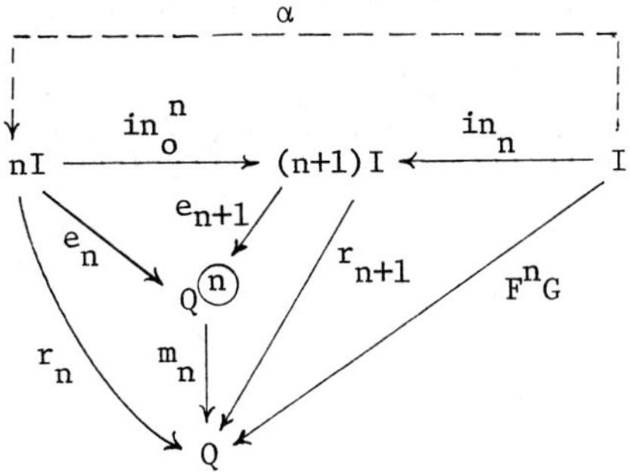


Noting the morphism $e_{n+1} \cdot in_n : I \to Q^{(n)}$ and the $\mathcal{E}$-character of $e_n : nI \to Q^{(n)}$, the projective property of I yields $\alpha: I \to nI$ with $e_n \cdot \alpha = e_{n+1} \cdot in_n$. Then $r_n \cdot \alpha = m_n \cdot e_n \cdot \alpha = m_n \cdot e_{n+1} \cdot in_n = r_{n+1} \cdot in_n = F^n G$. □

Various consequences follow from this theorem:

i) One can take duals (using injective (i.e. co-projective) output objects Y.

ii) One can take Q to be $Y_§$ and r to be $f^{\blacktriangle}$.

We then become interested in conditions on the so-called <u>Markov</u> parameters $A_i = HF^i G$, such as

$$A_n = \begin{pmatrix} A_0 \\ \vdots \\ A_{n-1} \end{pmatrix} \alpha , \qquad\qquad A_n = \theta\,(A_0 \;\ldots\; A_{n-1}).$$

(The column denoting a coproduct morphism, the row a product morphism.)

iii) Conditions such as $F^nG = r_n \cdot \alpha$ can even arise from conditions like

$$F^n = \begin{pmatrix} id \\ F \\ \vdots \\ F^{n-1} \end{pmatrix} \alpha .$$

Such a condition is fulfilled in Vect when F is finite-dimensional by the Cayley Hamilton Theorem, or in Grp when Q is finite (for then $F^N = id$ for some N).

All the theory to this point of the section has used the notion of $\mathcal{E}$-$\mathcal{M}$ factorization. Let us now change the viewpoint in order to study finite step conditions via the Nerode equivalence ideas.

4.15 ASSUMPTION: For the rest of this section the category $\mathcal{K}$ has countable coproducts and products. The process X of the machine $M = (X,Q,\delta,I,\tau,Y,\beta)$ is an adjoint process.

Recall from 2.7 that the abstract (see 2.1) and external (see 2.2) Nerode equivalences are the same as X is adjoint. We shall study conditions under which we can deal with only a finite number of the external equivalence conditions $f \cdot \alpha_n = f \cdot \gamma_n$ where the Nerode equivalence is $\alpha, \gamma \colon E_f \longrightarrow IX^@$, and $\alpha_n = \mu_0^{(n)} \cdot \alpha X^n$ is the n-step approximant to $\alpha^\#$.

4.16 DEFINITION: A pair of morphisms $\alpha, \gamma : E^N \longrightarrow IX^@$ are partially f-equivalent to level N if $f \cdot \alpha_n = f \cdot \gamma_n$, $n = 0,1,\ldots,N$ where, as usual, $\alpha_n$ denotes the n-step approximant to $\alpha^\#$. We say that $\alpha, \gamma : E_f^N \longrightarrow IX^@$ is a partial Nerode equivalence of f to level N if they are partially f-equivalent and if whenever $\bar{\alpha}, \bar{\gamma} : \bar{E}^N \longrightarrow IX^@$ are again partially f-equivalent, there exists a unique $\psi \colon \bar{E}^N \longrightarrow E_f^N$ such that

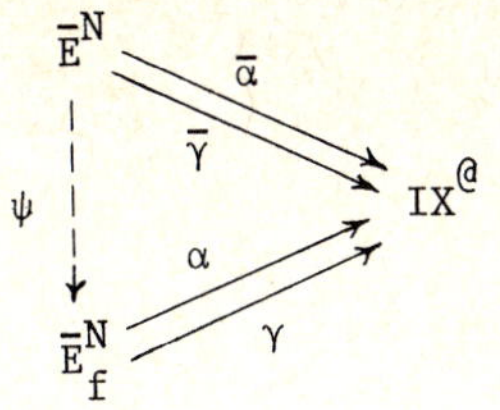


If $\mathcal{K}$ has kernel pairs, then as might be expected, $\alpha,\gamma$ is a kernel pair, actually of a type of approximant to $f^\blacktriangle$. We digress to establish this result.

<u>4.17</u> <u>DEFINITION</u>: The N-<u>step adjoint approximant</u> to $f^\blacktriangle$, denoted $f^\blacktriangle_N$, is obtained as follows. Let $[f\cdot\mu_0^{(n)}]^\cdot : \coprod_{j\geq 0} IX^j \to Y(X^\cdot)^n$ for $n\geq 0$ be the morphism obtained by the $X,X^\cdot$ adjunction from $f\cdot\mu_0^{(n)} : (\coprod_{j\geq 0} IX^j)X^n \to Y$. Then $f^\blacktriangle_N$ is defined by the product diagram:

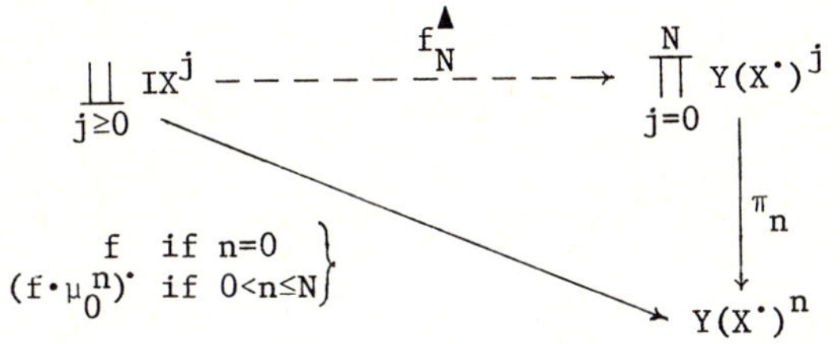


<u>4.18</u> <u>PROPOSITION</u>: The kernel pair of $f^\blacktriangle_N$ is the partial Nerode equivalence of f to level N in the sense that if either exists so does the other, and they are equal.

Proof: The result will follow from <u>4.16</u> and the kernel pair definition if for arbitrary $\bar{\alpha},\bar{\gamma}: \bar{E}^N \to IX^@$, the conditions $f^\blacktriangle_N\cdot\bar{\alpha} = f^\blacktriangle_N\cdot\bar{\gamma}$ and $f\cdot\bar{\alpha}_n = f\cdot\bar{\gamma}_n$, $0\leq n\leq N$ are equivalent. That this is so follows from the following equivalences:

$$f \cdot \bar{\alpha}_n = f \cdot \bar{\gamma}_n \qquad 0 \le n \le N$$

$$f \cdot \mu_o^{(n)} \cdot \bar{\alpha} X^n = f \cdot \mu_o^{(n)} \cdot \bar{\gamma} X^n \qquad 0 \le n \le N$$

$$\left.\begin{array}{ll} (f \cdot \mu_o^{(n)})^{\cdot} \cdot \bar{\alpha} = (f \cdot \mu_o^{(n)})^{\cdot} \cdot \bar{\gamma} & 0 < n \le N \\ f \cdot \bar{\alpha} = f \cdot \bar{\gamma} & \end{array}\right\} \quad \text{(use } X, X^{\cdot} \text{ adjointness)}$$

$$f_N^{\blacktriangle} \cdot \bar{\alpha} = f_N^{\blacktriangle} \cdot \bar{\gamma} \qquad \text{(use 4.17).} \qquad \square$$

Now we present the analogue of the morphisms established in 4.8.

4.19 THEOREM: Let $\alpha^N, \gamma^N : E_f^N \to IX^@$ and $\alpha^{N+1}, \gamma^{N+1} : E_f^{N+1} \to IX^@$ be partial Nerode equivalences to levels N and N+1 for a prescribed $f: IX^@ \to Y$. Then there exists a unique $s_{N+1}: E_f^{N+1} \to E_f^N$ and a unique $\ell_{N+1}: E_f^{N+1}X \to E_f^N$ :

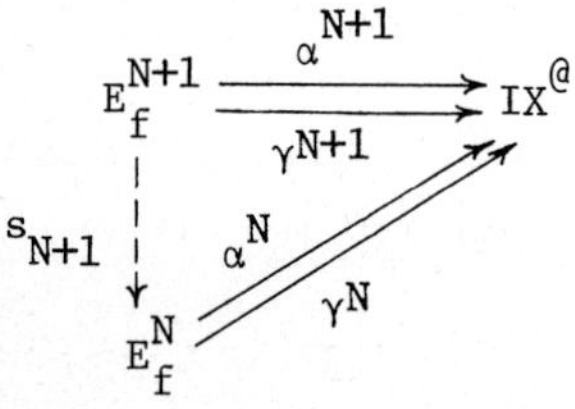


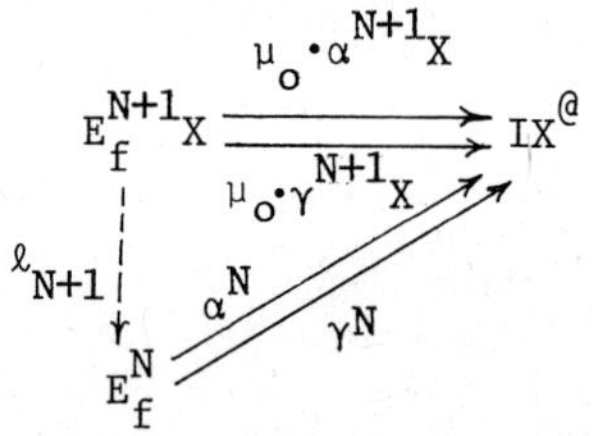


Proof: $f \cdot \alpha_n^{N+1} = f \cdot \gamma_n^{N+1}$ for $0 \le n \le N+1$ implies the same for $0 \le n \le N$. Identifying $\alpha^{N+1}$ and $\gamma^{N+1}$ with $\alpha$ and $\gamma$ in 4.16, $s_{N+1}$ exists (by identification with $\psi$ in 4.16). For $0 \le n \le N$, we have

$$\begin{aligned} f \cdot \mu_o^{(n)} \cdot (\mu_o \cdot \alpha^{N+1} X) X^n &= f \cdot \mu_o^{(n)} \cdot \mu_o X^n \cdot \alpha^{N+1} X^{n+1} \\ &= f \cdot \mu_o^{(n+1)} \cdot \alpha^{N+1} X^{n+1} && \text{(by definition of } \mu_o^{(n+1)}) \\ &= f \cdot \mu_o^{(n+1)} \cdot \gamma^{N+1} X^{N+1} && (E_f^{N+1} \text{ is a partial equivalence)} \\ &= f \cdot \mu_o^{(n)} \cdot (\mu_o \cdot \gamma^{N+1} X) X^N . \end{aligned}$$

Then we identify $\bar{\alpha}^N, \bar{\gamma}^N : \bar{E}^N \to IX^@$ in 4.16 with $\mu_o \cdot \alpha^{N+1}X$, $\mu_o \cdot \gamma^{N+1}X : E_f^{N+1}X \longrightarrow IX^@$ to conclude the existence of $\ell_{N+1}$. $\square$

The "when you stick, you're stuck" result follows:

4.20 THEOREM: With the definition and notation of 4.16 and 4.19, $s_{N+1}$ an isomorphism implies $s_{N+2}$ is an isomorphism and thus $s_{N+k}$ is an isomorphism for all $k > 2$.

Proof: First, let us show there exists $\phi\colon E_f^{N+1} \to E_f^{N+2}$. By 4.16, it is sufficient to show that $f\cdot\alpha_n^{N+1} = f\cdot\gamma_n^{N+1}$, $0 \le n \le N+2$. [Think of $\bar{E}^{N+2}$ as $E_f^{N+1}$.] Certainly this is true for $0 \le n \le N+1$. Now

$$\begin{aligned}
f\cdot\alpha_{N+2}^{N+1} &= f\cdot\mu_o^{(N+2)}\cdot\alpha^{N+1}X^{N+2} \\
&= f\cdot\mu_o^{(N+1)}\cdot\mu_o X^{N+1}\cdot(\alpha^{N+1}X)X^{N+1} \qquad \text{(by definition of } \mu_o^{(N+2)}\text{)} \\
&= f\cdot\mu_o^{(N+1)}\cdot(\mu_o\cdot\alpha^{N+1}X)X^{N+1} \\
&= f\cdot\mu_o^{(N+1)}\cdot(\alpha^{N}\cdot\ell_{N+1})X^{N+1} \qquad \text{(by 4.16)} \\
&= f\cdot\mu_o^{(N+1)}\cdot(\alpha^{N+1}\cdot(s^{N+1})^{-1}\cdot\ell_{N+1})X^{N+1} \\
&= (f\cdot\mu_o^{(N+1)}\cdot\alpha^{N+1}X^{N+1})\cdot((s^{N+1})^{-1})X^{N+1}.
\end{aligned}$$

Similarly, $f\cdot\gamma_{N+2}^{N+1} = (f\cdot\mu_o^{(N+1)}\cdot\gamma^{N+1}X^{N+1})\cdot((s^{N+1})^{-1})X^{N+1}$. Then $f\cdot\alpha_{N+2}^{N+1} = f\cdot\gamma_{N+2}^{N+1}$ since $f\cdot\mu_o^{N+1)}\cdot\alpha^{N+1}X^{N+1} = f\cdot\mu_o^{N+1)}\cdot\gamma^{N+1}X^{N+1}$.

At this stage, we have

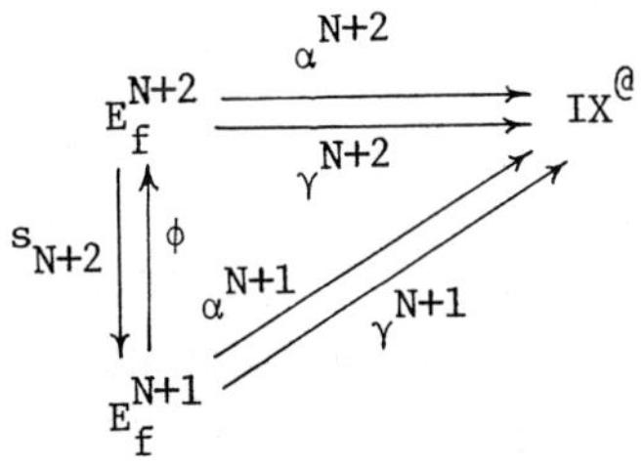


This diagram of itself does not show that $s_{N+2}$ is an isomorphism. However, by applying the partial Nerode equivalence definition to

$E_f^{N+2} \underset{\gamma^{N+2}}{\overset{\alpha^{N+2}}{\rightrightarrows}} IX^@$, it follows that both $id_{E_f^{N+2}}$ and $\phi \cdot s_{N+2}$ are candidates for the unique morphism $\psi\colon E_f^{N+2} = \bar{E}_f^{N+2} \longrightarrow E_f^{N+2}$ of definition 4.16. Thus $id_{E_f^{N+2}} = \phi \cdot s_{N+2}$ and likewise $s_{N+2} \cdot \phi = id_{E_f^{N+1}}$, so that $s_{N+2}$ is an isomorphism. □

Finally, and as one might expect, $\alpha^N, \gamma^N\colon E_f^N \longrightarrow IX^@$ can be taken as the external (and abstract, since X is adjoint) Nerode equivalence.

4.21 THEOREM: With the definition of 4.16, 4.19 and 4.20 and with $s_{N+1}$ an isomorphism, then $\alpha^N, \gamma^N\colon E_f^N \longrightarrow IX^@$ is the Nerode equivalence of f.

Proof: Use the isomorphism established in 4.20 to identify $E_f^N$, $E_f^{N+1}$, $E_f^{N+2}$, ... and $\alpha^N$, $\alpha^{N+1}$, $\alpha^{N+2}$, ... and $\gamma^N$, $\gamma^{N+1}$, $\gamma^{N+2}$, ... . Then $f \cdot \mu_o^{(n)} \cdot \alpha^N X^n = f \cdot \mu_o^{(n)} \cdot \gamma^N X^n$ for $0 \le n \le N$ by definition and for all $n > N$ by noting that $\alpha^n = \alpha^N$, $\gamma^n = \gamma^N$. Moreover, $f \cdot \mu_o^{(n)} \cdot \bar{\alpha} X^n = f \cdot \mu_o^{(n)} \cdot \bar{\gamma} X^n$ for all n implies this result for $0 \le n \le N$, whence there exists $\psi$ (by the partial Nerode equivalence property of $\alpha^N, \gamma^N$) such that

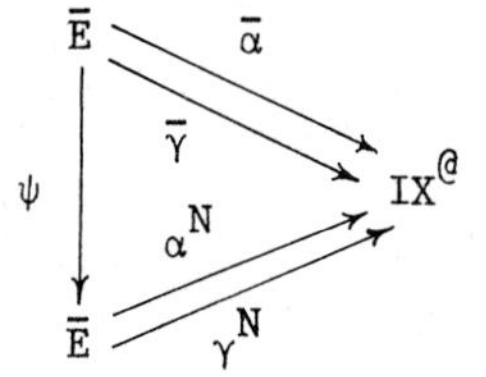


But this is precisely what is required to conclude the (non-partial) Nerode equivalence property of $E_f^N$. □

## 5. Augmenting the Process.

In section 4, we considered "application of less than n inputs" in terms of the coproduct $\coprod_{j=0}^{n-1} IX^k$, and then required X to preserve coproducts. Unfortunately, this requirement rules out tree automata--we know from [3] that the only $X: \underline{Set} \to \underline{Set}$ which preserves coproducts is the sequential machine process $X = - \times X_o$. In this section, we show how to achieve the goals of Section 4 in a less restrictive setting, by augmenting X to obtain a new functor $\bar{X}$

$$Q\bar{X} = QX + Q$$

(where + is the coproduct) whose n-fold application directly generates "application of at most n inputs". Consider the case $X = - \times X_o$. It is easy to check that $Q\bar{X}^n$ is then essentially $\coprod_{j \le n} (Q \times X_o^j)$, where the "essentially" reminds us that $Q \times X_o^j$ is actually present as $\binom{n}{j}$ disjoint copies. With this motivation, we turn to the general development:

5.1 DEFINITION: Let $X: \mathcal{K} \to \mathcal{K}$ where $\mathcal{K}$ has finite coproducts. Then we define the augmentation of X to be the process $\bar{X}: \mathcal{K} \to \mathcal{K} : Q \mapsto QX + Q$. Thus an $\bar{X}$-dynamics is just a pair

$$QX \xrightarrow{\delta} Q \quad \text{and} \quad Q \xrightarrow{F} Q$$

so that an $\bar{X}$-dynamorphism $h: (Q,\delta,F) \to (Q',\delta',F')$ must satisfy

$$\begin{array}{ccc} QX & \xrightarrow{\delta} & Q \\ {\scriptstyle hX}\downarrow & & \downarrow{\scriptstyle h} \\ Q'X & \xrightarrow{\delta'} & Q' \end{array} \qquad \text{and} \qquad \begin{array}{ccc} Q & \xrightarrow{F} & Q \\ {\scriptstyle h}\downarrow & & \downarrow{\scriptstyle h} \\ Q' & \xrightarrow{F'} & Q' \end{array}$$

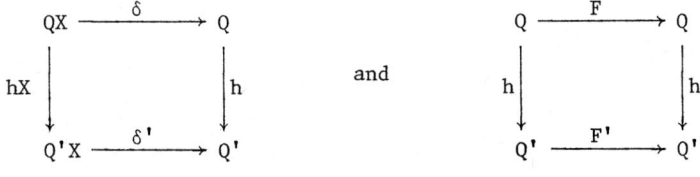

In applications, we shall be most interested in the case $F = id_Q$.

Where the theory of Section 4 only applied to processes which preserved finite coproducts, thus excluding the processes of tree automata, the current theory is not so restricted:

5.2 THEOREM: If X is an adjoint process or an $\Omega$-algebra process in Set, then $\bar{X}$ is an input process.

Proof: If X is an adjoint process, then $\bar{X}$ is an adjoint process

$$\frac{QX + Q \longrightarrow R}{Q \longrightarrow RX^{\cdot} \times R}$$

so that $\bar{X}$ has adjoint $\bar{X}^{\cdot}: R \longmapsto RX^{\cdot} \times R$ and so is an input process with $\bar{X}^{@} = \bar{X}^{*} = \coprod_{n \geq 0} \bar{X}^{n}$.

If X corresponds to the operator set $\Omega$, then $\bar{X}$ corresponds to the operator set $\bar{\Omega} = \Omega \cup \{1\}$ where $1 \notin \Omega$ and has arity 1. Thus $\bar{X}$ is an input process with $Q\bar{X}^{@}$ the free $\bar{\Omega}$-algebra on Q generators. □

We now retrace the stages of Section 4 in the augmentation setting. The results will not only have applicability to algebra automata, but will have somewhat simpler proofs:

Given an X-dynamics $\delta: QX \longrightarrow Q$ we define its augmentation to be the $\bar{X}$-dynamics

$$\bar{\delta} = \begin{pmatrix} \delta \\ Q \end{pmatrix} : QX + Q \longrightarrow Q.$$

Define the under-n-steps reachability map of $(\delta: QX \longrightarrow Q, \ \tau: I \longrightarrow Q)$ to be

$$\hat{r}_n = \begin{pmatrix} \delta \\ Q \end{pmatrix}^{(n-1)} \cdot \tau\bar{X}^{n-1} : I\bar{X}^{n-1} \longrightarrow Q.$$

Of course, $\hat{r}_n = \bar{r} \cdot \bar{\eta}_n$, where $\bar{r}$ is the $\bar{X}$-reachability map of $\bar{\delta}$.

It is then clear (observe that $I\bar{X}^n = I\bar{X}^{n-1}X + I\bar{X}^{n-1}$) that

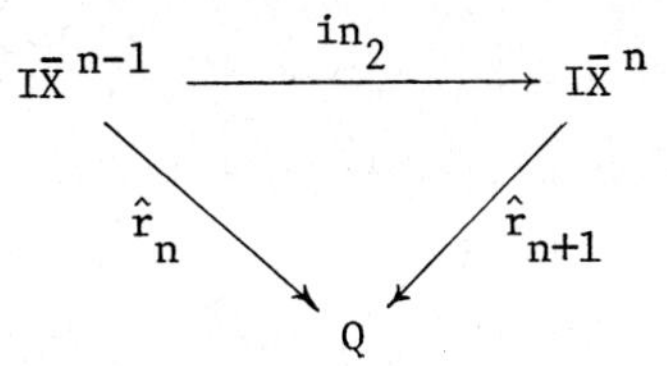


commutes. Analogously to 4.3 we then use diagonal fill-in to obtain

5.3 THEOREM: Let each $\hat{r}_n$ have an $\mathcal{E}$-$\mathcal{M}$ factorization $\hat{r}_n = \hat{e}_n \cdot \hat{m}_n$. Then there exists a unique $\hat{h}_{n+1}$ in $\mathcal{M}$ such that

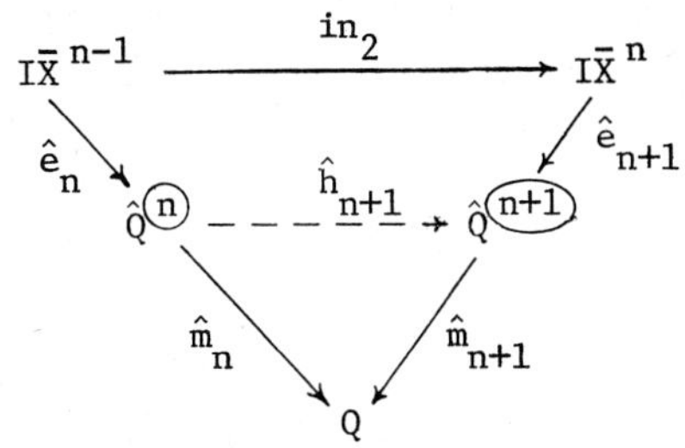


(1)

commutes. □

5.4 COROLLARY: If $\hat{r}_n \in \mathcal{E}$, then $\hat{r}_{n+1} \in \mathcal{E}$ and $\hat{Q}^{(n+k)} = Q$ for all $k \geq 0$. □

We now turn to the generalization of 4.5. We no longer require that X preserves finite coproducts (a restrictive condition that in Set forces $X = - \times X_0$) but only require that $\bar{X}$ preserves $\mathcal{E}$-morphisms (which in Set with $\mathcal{E}$ = onto --or in any category with $\mathcal{E}$ = split epimorphisms --is satisfied by every functor[†]):

[†] A split epimorphism $e: A \to B$ is one such that there exists $f: B \to A$ with $ef = id_B$. Note that not every category has $\mathcal{E}$-$\mathcal{M}$ factorization, and if a category has $\mathcal{E}$-$\mathcal{M}$ factorizations, it is not necessarily true that $\mathcal{E}$ = split epimorphisms is allowed.

<u>5.5</u> <u>THEOREM</u>: Let the augmented input process $\bar{X}$ preserve $\mathcal{E}$-morphisms. Then the following diagram defines a unique $\hat{k}_{n+1}: \hat{Q}^{\textcircled{n}}X \longrightarrow \hat{Q}^{\textcircled{n+1}}$.

$$\begin{array}{ccc}
(I\bar{X}^{n-1})\,\bar{X} & \xrightarrow{\quad id \quad} & I\bar{X}^{n} \\
{\scriptstyle \hat{e}_n\bar{X}}\downarrow & & \downarrow{\scriptstyle \hat{e}_{n+1}} \\
\hat{Q}^{\textcircled{n}}\bar{X} & \overset{\hat{k}_{n+1}}{\dashrightarrow} & \hat{Q}^{\textcircled{n+1}} \\
{\scriptstyle \hat{m}_n\bar{X}}\downarrow & & \downarrow{\scriptstyle \hat{m}_{n+1}} \\
Q\bar{X} & \xrightarrow{\quad \binom{\delta}{Q} \quad} & Q
\end{array} \tag{2}$$

Note that $\hat{k}_{n+1}$ is in $\mathcal{E}$.

Proof: That the outer square commutes is immediate from the definition of $\hat{r}_{n+1} = \binom{\delta}{Q} \cdot \hat{r}_n\bar{X}$. That $\hat{k}_{n+1}$ exists follows by the diagonal fill-in lemma, since $\hat{e}_n\bar{X} \in \mathcal{E}$. □

With this we come to the augmented "When you stick, you're stuck" theorem:

<u>5.6</u> <u>THEOREM</u>: Let the augmented input process $\bar{X}$ preserve $\mathcal{E}$-morphisms; and let $\hat{h}_{n+1}$ be defined as in <u>5.3</u>. If $\hat{h}_n$ is an isomorphism, then $\hat{h}_{n+1}$ is an isomorphism, and hence $\hat{h}_{n+k}$ is an isomorphism for all $k \geq 0$.

Proof: As in <u>4.6</u>, our problem reduces to showing that, when $\hat{h}_n$ is an isomorphism, then there exists $\hat{v}_{n+1}: I\bar{X}^n \longrightarrow Q^{\textcircled{n}}$ with $\hat{h}_{n+1} \cdot \hat{v}_{n+1} = \hat{e}_{n+1}$. But if $\hat{h}_n$ is an isomorphism, we may (compare (1) and (2)) set

$$\hat{v}_{n+1} = \hat{k}_n \cdot \hat{h}_n^{-1}\bar{X} \cdot \hat{e}_n\bar{X},$$

and we then observe that

$$\begin{aligned}
\hat{m}_{n+1} \cdot \hat{h}_{n+1} \cdot \hat{v}_{n+1} &= \hat{m}_n \cdot \hat{k}_n \cdot \hat{h}_n^{-1}\bar{X} \cdot \hat{e}_n\bar{X} && \text{by (1)} \\
&= \binom{\delta}{Q} \cdot \hat{m}_{n-1}\bar{X} \cdot \hat{h}_n^{-1}\bar{X} \cdot \hat{e}_n\bar{X} && \text{by (2)} \\
&= \binom{\delta}{Q} \cdot \hat{m}_n\bar{X} \cdot \hat{e}_n\bar{X} && \text{by (1)} \\
&= \binom{\delta}{Q} \cdot \hat{r}_n\bar{X} && \\
&= \hat{r}_{n+1} && \text{by definition of } \hat{r}_n \\
&= \hat{m}_{n+1} \cdot \hat{e}_{n+1}.
\end{aligned}$$

Thus, since $\hat{m}_{n+1} \in \mathcal{M}$, we have that $\hat{h}_{n+1} \cdot \hat{v}_{n+1} = \hat{e}_{n+1}$, as desired. □

Given the above isomorphism we may (analogously to the discussion preceding 4.7) set $\hat{Q} = \hat{Q}^n$, $\hat{k} = \hat{k}_{n+1}\colon \hat{Q}\bar{X} \to \hat{Q}$, $\hat{m} = m_n$ to obtain the diagram

$$\begin{array}{ccc}
\hat{Q}\bar{X} & \xrightarrow{\ \hat{k}\ } & \hat{Q} \\
{\scriptstyle \hat{m}\bar{X}}\downarrow & \binom{\delta}{Q} & \downarrow{\scriptstyle \hat{m}} \\
Q\bar{X} & \longrightarrow & Q
\end{array}$$

Setting $\hat{k} = \binom{\hat{\delta}}{F}$, we see that $\hat{m} \cdot F = \hat{m}$ and so $F = id_Q$, since $\hat{m} \in \mathcal{M}$. Thus $\hat{k}$ is indeed an augmented dynamics, based on $\hat{\delta}\colon \hat{Q}X \to \hat{Q}$. Our task, now, is to show that $\hat{\delta}$ yields the "reachable part" of $\delta$. The key, clearly, is to relate reachability via $I\bar{X}^{@}$ with reachability via $IX^{@}$. We can embed $I\bar{X}^{@}$ in $IX^{@}$ by the $\bar{X}$-dynamorphism $\bar{\psi}$ defined by:

$$\begin{array}{ccc}
I & \xrightarrow{\ \bar{\eta}\ } & I\bar{X}^{@} \\
& {\scriptstyle \eta}\searrow & \downarrow{\scriptstyle \bar{\psi}} \\
& & IX^{@}
\end{array}
\qquad\qquad
\begin{array}{ccc}
I\bar{X}^{@}\bar{X} & \xrightarrow{\ \bar{\mu}_o\ } & I\bar{X}^{@} \\
{\scriptstyle \bar{\psi}\bar{X}}\downarrow & \binom{\mu_o}{id} & \downarrow{\scriptstyle \bar{\psi}} \\
IX^{@}\bar{X} & \longrightarrow & IX^{@}
\end{array}
\qquad (3)$$

while $IX^@$ is sent into $I\bar{X}^@$ by the X-dynamorphism $\psi$ defined by

$$\begin{array}{ccc} I & \xrightarrow{\eta} & IX^@ \\ & {\scriptstyle\bar{\eta}}\searrow & \downarrow{\scriptstyle\psi} \\ & & I\bar{X}^@ \end{array} \qquad \begin{array}{ccc} IX^@X & \xrightarrow{\mu_o} & IX^@ \\ {\scriptstyle\psi X}\downarrow & & \downarrow{\scriptstyle\psi} \\ I\bar{X}^@X & \xrightarrow{in_1\cdot\mu_o} & I\bar{X}^@ \end{array} \tag{4}$$

The square of (3) tells us that

$$\begin{array}{ccc} I\bar{X}^@X & \xrightarrow{in_1\cdot\bar{\mu}_o} & I\bar{X}^@ \\ {\scriptstyle\bar{\psi} X}\downarrow & & \downarrow{\scriptstyle\bar{\psi}} \\ IX^@X & \xrightarrow{\mu_o} & IX^@ \end{array} \tag{5}$$

so that $\bar{\psi}$ is also an X-dynamorphism. Moreover, splicing the diagrams of (4) above those of (5), we deduce that

$$\bar{\psi}\cdot\psi = id_{IX^@}.$$

[On the other hand, the square of (4) yields

$$\begin{array}{ccc} IX^@\bar{X} & \xrightarrow{\binom{\mu_o}{id}} & IX^@ \\ {\scriptstyle\psi\bar{X}}\downarrow & & \downarrow{\scriptstyle\psi} \\ I\bar{X}^@\bar{X} & \xrightarrow{\binom{\bar{\mu}_o\cdot in_1}{id}} & I\bar{X}^@ \end{array}$$

This will imply $\psi\cdot\bar{\psi} = id_{I\bar{X}^@}$ only if it is the case that $\bar{\mu}_o = \begin{pmatrix} \bar{\mu}_o\cdot in_1 \\ id \end{pmatrix}$, but this case is too restrictive to detain us here.]

Now by the definition of $\hat{Q}$, we may factor $\tau$ through $\hat{Q}$, say

$$\tau = I \xrightarrow{\hat{\tau}} \hat{Q} \xrightarrow{\hat{m}} Q.$$

Consider, then, the $\bar{X}$-dynamorphic extension of $\tau$:

$$\begin{array}{ccc} I & \xrightarrow{\bar{\eta}} & IX^@ \\ & {\scriptstyle\hat{\tau}}\searrow & \downarrow{\scriptstyle r} \\ & {\scriptstyle\tau}\searrow & \hat{Q} \\ & & \downarrow{\scriptstyle\hat{m}} \\ & & Q \end{array} \qquad \begin{array}{ccc} I\bar{X}^@\bar{X} & \xrightarrow{\bar{\mu}_o} & I\bar{X}^@ \\ {\scriptstyle\tilde{r}\bar{X}}\downarrow & & \downarrow{\scriptstyle\tilde{r}} \\ \hat{Q}\bar{X} & \xrightarrow{\hat{k}} & \hat{Q} \\ {\scriptstyle\hat{m}\bar{X}}\downarrow & & \downarrow{\scriptstyle\hat{m}} \\ Q\bar{X} & \xrightarrow{\binom{\delta}{Q}} & Q \end{array}$$

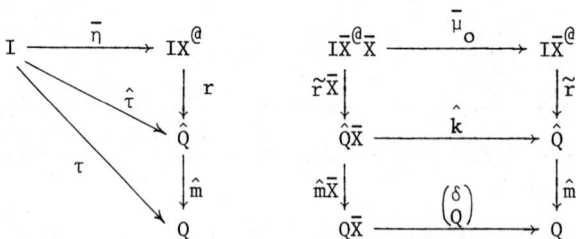

with the right-hand rectangles yielding

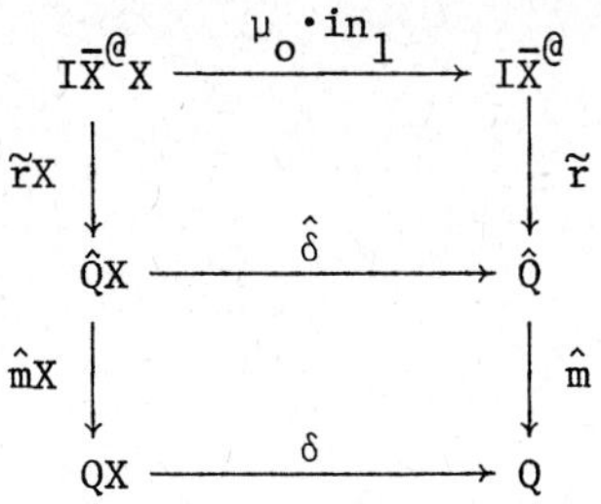


Splicing (4) atop this we see that $\hat{r} = \tilde{r}\psi$ is the X-reachability map of $\hat{\delta}$,

and that

$$r = \hat{m} \cdot \hat{r}.$$

Since $\hat{m}\cdot\tilde{r}\cdot\bar{\eta}_n = \hat{r}_n = \hat{m}\cdot\hat{e}_n$, we have $\tilde{r}\cdot\bar{\eta}_n = \hat{e}_n \in \mathcal{E}$, and so $\tilde{r} \in \mathcal{E}$.

We saw above that $\hat{m}\cdot\tilde{r}$ is the $\bar{X}$-dynamorphic extension of $\tau$. But so too is $r\cdot\bar{\psi}$, in view of the diagram

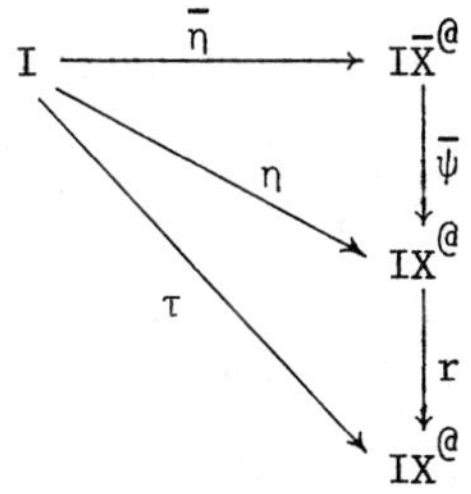


Thus we have $r\cdot\bar{\psi} = \hat{m}\cdot\tilde{r}$, and so we may let $\mathring{m}\cdot\mathring{e}$ be an $(\mathcal{E},\mathcal{M})$-factorization of r to form the diagram:

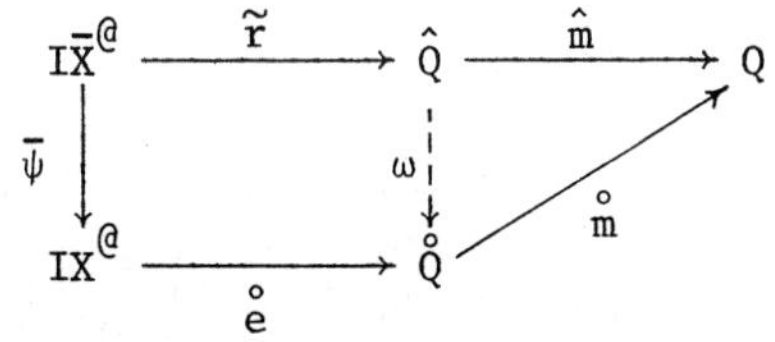


Recalling that $\bar{\psi}\cdot\psi = id_{IX^@}$, we see that $\bar{\psi} \in \mathcal{E}$, and hence $(\bar{\psi}\mathring{e})\cdot\mathring{m}$ is another $\mathcal{E}$-$\mathcal{M}$ factorization of $\hat{r}$. Thus there is an isomorphism $\omega\colon \hat{Q} \to \mathring{Q}$

such that

$$\omega \cdot \tilde{r} = \mathring{e} \cdot \bar{\psi} \quad \text{so that} \quad \omega \cdot \tilde{r} \cdot \psi = \mathring{e},$$

so that $\tilde{r} \cdot \psi = \omega^{-1} \cdot \mathring{e} \in \mathcal{E}$.

Thus we may use the $\mathcal{E}$-$\mathcal{M}$ factorization

$$r = IX^{@} \xrightarrow{\tilde{r}\cdot\psi} \hat{Q} \xrightarrow{\hat{m}} Q.$$

We have proved the following:

5.7 THEOREM: Let X be an input process whose augmented process is an input process $\bar{X}$ which preserves $\mathcal{E}$-morphisms. If $\hat{h}_n$ as defined in 5.3 for the pair $(\delta: QX \to Q,\ \tau: I \to Q)$ is an isomorphism, then setting $\hat{m} = \hat{h}_{n-1}$, $\hat{\delta} = \hat{k}_n$ and $\hat{\tau} = h_n \cdot \ldots \cdot h_o \cdot \hat{e}_o$, we have that $(\hat{\delta}: \hat{Q}X \to \hat{Q},\ \hat{\tau}: I \to \hat{Q})$ is reachable, with reachability map $\hat{r}: IX^{@} \to \hat{Q}$ in $\mathcal{E}$ satisfying $\hat{m} \cdot \hat{r} = r$. That is, $(\hat{\delta},\hat{\tau})$ is the "reachable part" of $(\delta,\tau)$. □

Note that this does not say that $(\hat{\delta},\hat{\tau})$ is reachable in any finite-number of steps using the unaugmented input process--compare the comments following 3.12.

We shall not belabor the observability story in this section, since the $\Omega$-algebra process does not have an observability theory [3], save to note that the dual of the augmentation of an adjoint process is the augmentation of its dual:

$$\frac{\dfrac{QX + Q \longrightarrow R}{Q \longrightarrow RX^{\cdot} \times R}}{RX^{\cdot} +_{op} R \longrightarrow\!\!< Q} \quad : \quad \overline{(X^{\cdot})^{op}} = (\bar{X}^{\cdot})^{op}.$$

## CONCLUSIONS

To summarize the preceding material in a short space, we claim three major ideas:

1) The application of $\mathcal{E}$-$\mathcal{M}$ factorizations in a category to the realization problem;

2) The definition of a Nerode equivalence within a category, and its application to the realization problem;

3) The formulation of the finite-dimensionality idea of linear systems theory and sequential machines in a category theory framework, and its application to machine realization.

In the remainder of these conclusions, we shall discuss some possible criticisms of the theory. At the outset, let it be seen that we do not claim that the approach of this paper should supplant other and more traditional approaches. Rather, we would claim the advantages noted in the introduction, of unification and exposure of the essentials of some results. [Nor, indeed, would a category theorist claim that the study of group theory could be supplanted by the study of category theory, despite the fact that there is a category of groups.]

In rebuttal to the positive claims concerning the material of the paper, we could argue as follows:

1) The unification may be more illusory than real. To say that state behavior machines unify adjoint machines is not that helpful, when we believe that the right way to study adjoint machines is as adjoint machines, rather than as state-behavior machines. What then is the value of state-behavior machines, since we know of no examples (other than contrived ones) of such machines which are not also adjoint machines. [As partial defense against this attack, we can claim that decomposable machines, adjoint machines, and input process machines each include more than one example of stature--linear

systems and group machines, sequential machines and Goguen's affine machines [13], tree automata and non-deterministic automata [1], respectively.]

2) The unification comes only with the aid of concepts which many, including engineers at least, find very hard to master. Though the general theory of relativity may unify special relativity and Newtonian mechanics, engineers prefer to leave the general theory of relativity to the physicists, and stick with the Newtonian mechanics. Likewise, they may wish to leave the category theory to category theorists. [As partial defense, we could claim that the ideas of category theory used here are the more elementary ideas of the subject, and that the subject would itself be much easier to learn if a textbook written assuming less advanced knowledge of the reader had been available; that such a defect has now been remedied is the contention of the authors [12, 20]. And we might also argue that the linear algebra now regarded as both commonplace in and essential for the study of control theory was once viewed by engineers as being alarmingly difficult.]

3) As yet, the theory has done little new for the control theorist remaining within the confines of linear systems, or finite-dimensional non-linear systems. Nor has it shown how to even formulate in a category theory framework that most fundamental of control theory notions--feedback. The search for such a formulation is a major challenge for those who would bring category theory to the attention of control theorists. On the other hand, Goguen's affine machines [13] model a discretized system ((1) of Chapter 1) which is neither initialized nor fully linearized as a bilinear term in q and u is retained. As adjoint machines, the realization theory is clear. However, an element of the "object of inputs" $IX^@$ is more complex than a finite sequence of inputs. Here, the philosophy of category theory suggests a new principle in system engineering: for nonlinear systems, the structure

of "input strings" is not dictated by a priori intuition. A suitable algebraic theory of discretized nonlinear systems (which does not exist at this writing) may be due to the failure to recognize the proper formulation of the response $f: IX^{@} \rightarrow Y$ of such systems.

## REFERENCES


1. M.A. Arbib and E.G. Manes (1974), Machines in a Category: An Expository Introduction, SIAM Review, 16, 163-192.

2. M.A. Arbib and E.G. Manes (1974), Foundations of System Theory: Decomposable Systems, Automatica, 10, 285-302.

3. M.A. Arbib and E.G. Manes (1975), Adjoint Machines, State-behavior machines, and Duality, J. Pure Appl. Algebra, in press.

4. R.E. Kalman (1969), Algebraic Theory of Linear Systems, in Topics in Mathematical System Theory (by R.E. Kalman, P.L. Falb, and M.A. Arbib), McGraw-Hill (1969).

5. R.W. Brockett and A.S. Willsky (1973), Finite-State Homomorphic Sequential Machines, IEEE Trans. Aut. Control, AC-17, 483-490.

6. M.A. Arbib (1973), Coproducts and Group Machines, J. Comp. Syst. Sci., 7, 278-287.

7. L.S. Bobrow and M.A. Arbib (1974), Discrete Mathematics: Applied Algebra for Computer and Information Science, Washington: Hemisphere Publishers.

8. A. Nerode (1958), Linear Automaton Transformations, Proc. Amer. Math. Soc., 9, 541-544.

9. M.A. Arbib and H.P. Zeiger (1969), On the Relevance of Abstract Algebra to Control Theory, Automatica, 5, 589-606.

10. S. Mac Lane (1972), Categories for the Working Mathematician, Springer-Verlag.

11. H. Herrlich and G.E. Strecker (1973), Category Theory, Boston: Allyn and Bacon.

12. M.A. Arbib and E.G. Manes (1975), Arrows, Structures, and Functors: The Categorical Imperative, New York: Academic Press.

13. J.A. Goguen (1972), Minimal Realizations of Machines in Closed Categories, Bull. Amer. Math. Soc., 78, 777-783.

14. H. Ehrig and J. Kreowski (1974), Power and Initial Automata in Pseudo-closed Categories, in Category Theory Applied to Computation and Control (E.G. Manes, ed.), Lecture Notes in Computer Science 25, Springer-Verlag, 144-150.

15. P.M. Cohn (1965), Universal Algebra, New York: Harper and Row.

16. G. Grätzer (1967), Universal Algebra, Princeton, N.J.: Van Nostrand.

17. Y. Give'on and M.A. Arbib (1968), Algebra Automata II: The Categorical Framework for Dynamic Analysis, Inform. Control, **12**, 346-370.

18. E.G. Manes, Algebraic Theories, Graduate Texts in Mathematics **26**, Berlin: Springer-Verlag, in press.

19. M.A. Arbib and E.G. Manes (1974), Time-Varying Systems, in Category Theory Applied to Computation and Control, Lecture Notes in Computer Science **25**, Springer-Verlag, 87-92.

20. L. Padulo and M.A. Arbib (1974), System Theory: A Unified State-Space Approach to Continuous and Discrete Systems, Washington: Hemisphere Publishers.

21. E.S. Bainbridge (1972), A Unified Minimal Realization Theory, with Duality, for Machines in a Hyperdoctrine, dissertation, University of Michigan.

22. J. Adámek (1974), Free Algebras and Automata Realizations in the Language of Categories, Comm. Math. Univ. Carolinae, **15**, 589-602.

23. V. Trnková (1974), On Minimal Realizations of Behavior Maps in Categorical Automata Theory, Comm. Math. Univ. Carolinae, **15**, 555-566.

24. V. Trnková (1975), Minimal Realizations for Finite Sets in Categorical Automata Theory, Comm. Math. Univ. Carolinae, **16**, 21-35.